D0792173

Mobile Web Services

Mobile Web services are designed to provide access to Web content anywhere and anytime, and often on any device. This book describes the key network elements, software components, and software protocols that are needed to realize these services, including the concept of user context and its potential to create personalized services. Major functions needed to implement the wireless mobile Web are explained in detail and cover location representation and tracking, security schemes, content personalization approaches, privacy mechanisms, and XSLT processing for browser content generation. WAP and i-mode mobile network architectures are examined. The author reviews latest mobile phone features and describes key aspects of browser markup languages (WML, cHTML, and XHTML MP). Ontology concepts and their application to enable the wireless Semantic Web are described and this book puts forward a novel definition and categorization of mobile user context with a detailed specification of the context ontology in W3C's Resource Description Framework (RDF) Schema. A mobile network architecture is presented, with in-depth explanation of each function, software infrastructure, and communication protocols (including SOAP, and the related WSDL) and an elaborated case study with code samples in XML and Java is included. The book is intended for wireless Web architects, network managers, and graduate students in electrical engineering and computer science.

Ariel Pashtan is president of Aware Networks Inc., developing Web-based services and enabling technologies for the wireless Internet. He has over 20 years of software research-and-development experience working for Motorola, IBM, Gould, and Israel Aircraft Industries. His academic experience includes appointments at Northwestern University, Evanston, Illinois, and at the Technion, Israel Institute of Technology, Haifa, Israel; directing students' research projects; and teaching operating systems courses.

Mobile Web Services

Ariel Pashtan

Buffalo Grove, IL, USA

CAMBRIDGE UNIVERSITY PRESS
Cambridge, New York, Melbourne, Madrid, Cape Town, Singapore, São Paulo

Cambridge University Press
The Edinburgh Building, Cambridge CB2 2RU, UK

www.cambridge.org
Information on this title: www.cambridge.org/9780521830494

First published 2005

Printed in the United Kingdom at the University Press, Cambridge

A catalog record for this book is available from the British Library

Library of Congress Cataloging in Publication data
Pashtan, Ariel.
Mobile Web services/Ariel Pashtan.
p. cm.
Includes bibliographical references and index.
ISBN 0 521 83049 4
1. Wireless communication systems. 2. Web services. I. Title.
TK5103.2.P38 2004 004.67′8 – dc22 2004054623

ISBN-13 978-0-521-83049-4 hardback
ISBN-10 0-521-83049-4 hardback

To Itai, Elad, and Tamar.

Contents

Figures

Preface

The future mobile Web

The "always on" vision of mobile Internet access has become a reality with the nearly ubiquitous coverage provided by cellular networks. Communication speeds have also increased significantly with the advent of third generation (3G) cellular networks that enable data transfer rates supportive of real-time video. In addition, competing access technologies such as wireless local area networks provide hotspots where a wireless user can experience Internet data rates similar to these available with broadband connectivity in the fixed network.

Over 100 million wireless Internet users were recorded as of September 2003, with the majority in Japan and Korea, while fast growth rates were experienced in Europe. Significant growth is expected in specialized mobile services such as driving directions, traffic reports, tour guides, and commerce services such as mobile shopping. To use these services, the mobile terminal plays the role of both a "network computer" for retrieving relevant information and a "wallet PC" for enacting related transactions. Users could use their mobile terminal to search and order products to be delivered at a nearby store, and, once at the store, they could use e-money to pay for the products with their mobile terminal.

The present book, as its title "Mobile Web Services" suggests, describes the building blocks needed to put together mobile networks that can wirelessly deliver Web content. Included is extensive coverage of the network elements, languages used to represent browser content, communication protocols, network services, and related software components that are used in the operation of such networks. The described network services include user location tracking, security schemes, content personalization approaches, privacy mechanisms, and style sheet processing for browser content generation.

While on the move, a mobile user faces many challenges such as a mobile terminal's limited screen size, restricted input capabilities, battery power constraints, and air time costs. This is where knowledge of a user's context can be leveraged to drive and personalize the interaction between user and Internet server so as to ease the communication exchange and focus the delivery of Web content to information that is directly applicable to a user's situation. This book presents scenarios of mobile users that seek Web content and describes the pertinent elements of context that affect content delivery. A corresponding ontology of user and environment context and its formal representation in the W3C's Resource Description Framework (RDF) Schema is elaborated upon.

Context-aware Web access requires the support of network services that are realized in a mobile Web network architecture whose network element functions, message flows, and system software support are explained. The book concludes with a description of an experimental mobile Web network. This network delivers tourism information to mobile users in a city environment and the system description includes related code samples in XML and Java.

Readers of "Mobile Web Services" include mobile Internet engineering managers, system architects, software developers, and engineering students wishing to learn how to design mobile Web networks.

How this book is organized

Chapter 1: *The mobile Web landscape* introduces the concept of mobile Web services and the mobile Web networks that provide the framework where mobile Web services can be deployed. A description of context, a key element in the realization of mobile services follows. Finally, this chapter lists the standards organizations that contribute mobile Web-related specifications, and concludes with an introduction to XML, the language used in most program examples provided in this book.

Chapter 2: *Wireless system architecture* explains the wireless system architecture specified by the WAP Forum (now the Open Mobile Alliance). A description of NTT DoCoMo's i-mode wireless architecture follows. This chapter describes how both the WAP and i-mode architectures have converged to a wireless Internet system, and shows how these architectures are realized over GSM's GPRS data network.

Chapter 3: *Wireless terminals and wireless content* reviews some of the most recent mobile phone capabilities and includes a description of browser-based applications, and the associated markup languages used in the creation of wireless content: WML, Compact HTML, and XHTML Mobile Profile. The chapter concludes with an elaboration of the evolution of browser capabilities.

Chapter 4: *User mobility and location management* starts with a review of IP addressability and the mobile IP standard followed by a description of handset-based and network-based approaches for locating mobile users. Tree-based hierarchical schemes for representing location and spatial information are described next, with an ensuing description of moving objects databases for storing spatial-temporal information. The chapter concludes with a description of the US E911 emergency services solutions for locating mobile users in distress.

Chapter 5: *Wireless network security* reviews the objectives of security for mobile environments. This chapter describes common methods for securing the transmission of messages. It proceeds with a depiction of mobile terminal and server authentication, and layouts an authorization framework. This chapter describes in detail Web services security as specified by the W3C and the OASIS standards organization.

Chapter 6: *Personalization and privacy* introduces the benefits of personalization of mobile interactions. Approaches used to build user behavior models are reviewed, followed by an elaboration of recommenders that tailor Web information delivered to mobile users. Included is a description of the architectural components needed to realize a personalization system. This chapter also addresses privacy concerns and details the related W3C's P3P effort.

Chapter 7: *Ontologies and RDF Schema* reviews ontology concepts and their application to enable the wireless Semantic Web. Efforts related to mobile services in the W3C and FIPA standards organizations are described. Approaches and criteria used in the generation of ontologies are elaborated upon next. An introduction to the W3C-defined RDF and RDF Schema (RDFS) are introduced next, as they provide a formal framework for defining an ontology. This chapter concludes with a description of the evolution of Web ontology languages that build on RDFS.

Chapter 8: *Ontology of mobile user context* provides a motivation for the definition of a mobile user's context. Four major scenarios of wireless Internet access by mobile users that wish to access Web-based content are presented. For each scenario, the activity of a corresponding context-aware service that delivers Web content is elaborated upon. Associated static and dynamic context elements are described and presented in the form of RDFS graphs and RDFS code.

Chapter 9: *XSLT for Web content presentation* starts by explaining how to generate XML representations of Web-based content that take into account user context. A description follows on how an XSLT processor can apply XSLT style sheets to XML representations to generate browser markup. Core capabilities of XSLT style sheet programming are introduced, and it is shown how to leverage XSLT for generating displayable content that reflects awareness of a mobile user's situation.

Chapter 10: *Mobile Web network* layouts the architecture of a network that includes a context manager and enables the delivery of mobile Web information. The functions of each network element, as well as message flows, are elaborated upon. All fixed network elements are implemented as Web services, and this chapter provides a detailed description of the underlying W3C SOAP communication protocol and WSDL Web services interface specification.

Chapter 11: *Context-aware tourist information system* elaborates on an experimental mobile Web network referred to as CATIS. CATIS implements a context-aware architecture that delivers tourism information to mobile users in a city environment. This chapter describes the network's architecture, network element functions, message flows, system interfaces, and managed content. Included are detailed code samples in XML and Java.

Acknowledgments

I gratefully acknowledge the inputs, reviews, and comments of my colleagues Dale Buccholz and Peter Scheuermann. Candace Locklear and Vanessa Veneziano from Openwave Systems were very supportive and provided review comments from Openwave's Client Product Group.

Andi Heusser and Remy Bläettler, former graduate students of Northwestern University whose projects I advised, helped develop and test systems and algorithms for mobile Web services, and contributed the code samples in the chapter describing a context-aware tourist information system. I am grateful for their enthusiasm and dedication to explore this exciting area.

I thank my former Motorola Labs colleagues for the many wireless user experience insights learned from the experimental Motorola Museum wireless tour guide.

The Cambridge University Press staff were very encouraging, and I thank Phil Meyler and Emily Yossarian for their assistance and guidance throughout the writing endeavour.

My appreciation and thanks also to Alex Sharpe for the editing, and to Jayne Aldhouse from the production team.

Finally, this book was completed thanks to the commitment and encouragement of my wife Hanna, my children Itai, Elad, and Tamar, and my parents Dov and Rachel.

1 The mobile Web landscape

Mobile Web services are introduced in this chapter. We describe the type of services that mobile users are likely to invoke when connected to the wireless Internet and the browser-based access that characterizes how these services are most often used. We proceed with a description of the associated service delivery models that explain how these services can be developed and deployed. There is significant evidence from the NTT DoCoMo network operator experience that shows the success of an "open" model where the mobile operator provides a framework and environment in which third party content developers can deploy their services. The major stakeholders in the wireless Internet ecosystem are examined next. They include the mobile user, the enterprise, the mobile network operator, the content provider, and finally the mobile network and handset manufacturer.

To support the "open" service development model, the underlying networks need to provide a set of services that are easily accessed by wireless service developers. We describe the concept of a mobile network operating system and the value that Web services technology can bring to bear in the realization of this model. We proceed with an introduction of the concept of context, elaborate on its potential to create personalized services, review the pertinent elements of context, and describe how context is collected and stored in the network. Many standard bodies are contributing specifications that will enable the realization of the mobile Web, and we list the major organizations and their stated objectives. Finally, we provide an introduction to XML and XML Schema, the languages used in most of the example code listings in this book.

1.1 Mobile Web services

The currently deployed new generations of mobile networks (2.5G and 3G) that support data services provide an always-on connect capability, with typical data transfer rates of about 50 kilobits per second (kbps) for 2.5G and about 144 kbps and higher for 3G, and, most importantly, their user pricing models are either based on volume of sent packets or else on a flat monthly fee. These mobile networks extend the wired Internet and allow users to have access "anytime and anywhere" to the same information that they are relying on in their home or office. This is a significant change from previous generation mobile networks that were circuit switched with dial-in access, provided low data transfer speeds (typically 9.6 or 14 kbps), and required users to pay a "per minute" connect time fee.

With the wireless Internet, unique information services can be offered which were not available before. Defined user groups with very specific information needs can be easily accessed in real time. For example, music buffs can receive on their cell phone notification from their music store about the current availability of their favorite music CDs, they can reserve a copy of a CD, and they can pay for them with their phone. Similarly, there are surfer sites (for example, Surfline [1]) that can provide up-to-date information on surf conditions along the coast and enable surfers to head to the right beach. The Surfline site provides streaming video of various coasts in the USA and other places in the world. These streams could be sent to mobile users equipped with phones capable of displaying video so that they can decide for themselves whether the surf conditions are optimal.

Mobile Web services address a wide gamut of user interests. These services can be partitioned into distinct categories, some examples of which are:

- General information
  - news feeds;
  - stock quotes;
  - weather reports;
  - horoscopes;
  - online dictionary.
- Travel
  - car navigation;
  - traffic reports;
  - train connections check;
  - airline ticket reservation;
  - flight status check.
- Entertainment
  - restaurant reservation;
  - movie theater reservation;
  - games.
- ecommerce
  - mobile shopping;
  - online banking.
- Connectivity
  - email;
  - finding friends.

Access to these services and their associated content is typically performed through a Web browser. On the wired Internet, information access is performed via Web browsers such as Microsoft's Internet Explorer and Netscape's Navigator that are able to display content that has been marked up with formatting tags. Hypertext Markup Language (HTML) is the primary markup language used to annotate Internet content for browser display. On the wireless Internet, two major browsers used in mobile terminals include the Wireless Application Protocol (WAP) browser specified by the Open Mobile Alliance (OMA) and the i-mode browser specified by NTT DoCoMo. The WAP browser can display mobile content in Wireless Markup Language (WML) and Extensible HTML (XHTML), and the i-mode browser, used mainly in Japan, can display content in Compact HTML (cHTML), a subset of HTML. Dual mode browsers enable access to both WAP and i-mode content.

Over 400 million handsets support the Openwave Systems WAP browser as of July 2003 [2]. Openwave shipped the world's first WAP 2.0 browser during 2001 in Japan, where it is the second most popular browser. Successful mobile data services such as Sprint PCS Vision, KDDI "au", and Vodafone live! are based on WAP technology. Over 20 million Vodaphone live! subscribers, as of July 2004, were using a WAP browser while Japan's DoCoMo had over 40 million subscribers, as of October 2003, accessing the Internet via an i-mode browser [3]. Besides the Japanese handset manufacturers such as NEC, Panasonic, and Mitsubishi, as of 2003 other manufacturers such as Samsung (DoCoMo phones), Siemens (S55 mobile phone), and Nokia (3650 handset), include an i-mode browser in select mobile phones.

As in the wired Internet, the wireless user can leverage a search engine to find relevant services and content within the available wireless Web pages. For example, Google has a wireless search service [4] that is offered by wireless operators that include Sprint, AT&T Wireless, Cingular, Qwest, Nextel, Vodafone, and others. Google's search technology can be accessed from any number of terminals, including mobile phones, PDAs, and the Pocket PC. The Google service searches through the over five million WML pages created for wireless WAP terminals. It can also expand its search to the over three billion pages of the World Wide Web. Every viewed page is translated on the fly to text and all search results are displayed in text-only format that fits the terminal screen. The i-mode service has similar search engines, for example, to answer user queries the Oh!New? engine will search Compact HTML pages in i-mode sites.

1.2 Mobile Web service delivery

With the breadth of information and service types that wireless users wish to access, no mobile operator can provide all the required solutions. In the old telecom model, the operator was responsible for providing network communication features and these were rolled out in the network in successive versions of the network software. In voice networks, these features included facilities such as call forwarding, and voice mail.

This was a network-centric model where intelligence was provided in the operator's network servers.

In the Internet model, intelligence has migrated to the edge of the network, and is provided mainly by third parties that specialize in specific content offerings. Network operators could adopt the "transport system" model, where they don't own or manage content and only provide transport services. The drawback of this model is that communication prices become the only tool for customer retention, putting significant pressure on operator revenues. An alternative model is for the operator to take responsibility to develop, aggregate, and deliver mobile content. The operator's portal site links to a select set of third party content providers with whom the operator has established licensing agreements. In this "closed" environment, the operator has sole responsibility for deciding what content is offered on its portal, and the content providers' customers are the operators, not the end-user subscribers. This model was adopted until recently by most operators outside of Japan [5]. The evident drawback of this model is that no incentives were in place for content providers to develop new and exciting services, as they had no way to charge the end user for any improvements. In some cases, the content providers had to pay the operator for an entry on its wireless portal site. Mobile subscribers complained about the lack of variety in services, which in turn led to limited service use, so that ultimately the operators experienced a limited growth in wireless Internet usage.

Mobile operators such as NTT DoCoMo, have made conscious decisions to become a coordinator and facilitator for service delivery and not to perform Web content development or even not to purchase content from Internet companies [6]. DoCoMo chose to become a portal site operator and facilitate user navigation to find useful content. The i-mode portal menus contained 3600 sites as of June 2003. DoCoMo established a very elaborate procedure for evaluating content providers and deciding which would appear on the official i-mode portal. This role is not unlike the one of wired Internet providers such as AOL. AOL too, does not create content but rather provides a platform for content delivery which is accessed via its portal. In this latter service model, the operator provides a platform, the wireless network, for service delivery and service management functions (see Figure 1.1). Content providers can rely on the billing services provided by the operator to track usage and bill the users. A corresponding fee is paid to the operator; for example, a nine percent fee is levied in the case of DoCoMo from subscriber fees paid to content providers.

While mobile operators have chosen to provide official site menus, they allow users to access the broader Internet as well. The user has to key in the Universal Resource Locator (URL) of a specific site he/she wishes to reach: not a very convenient task on a phone device. Nevertheless, it is interesting to note that the number of sites that are accessed outside the operator's portal far exceeds the number of operator provided sites. DoCoMo counted 66 400 i-mode sites, as of June 2003, which were not listed in its menus. These sites usually provide dedicated content for a small number of users that would not justify a slot on i-mode's official menu. The mobile service provider's

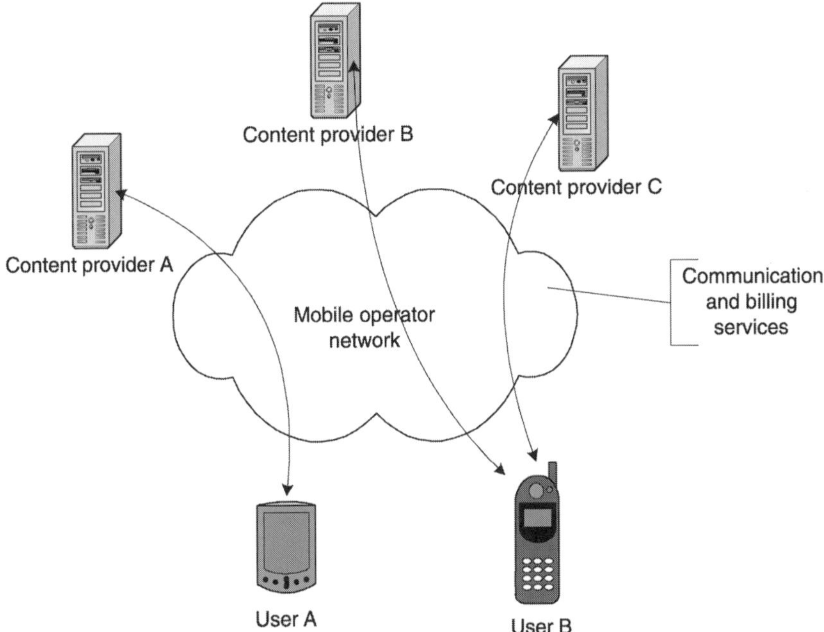

Figure 1.1 Mobile operator: bridging between content providers and mobile users.

revenue from wireless communication charges to these unofficial sites can be very significant and even exceed the average revenue per user (ARPU) from official sites, as in the case of DoCoMo [5].

Early success of wireless Internet use occurred in Japan. The penetration of cellular phones is higher than that of personal computers, one of the reasons being the high access cost to wired Internet services. The cost of DoCoMo's i-mode wireless Internet access is by contrast a small fee (about \$3) per month, a content service subscription fee (between \$1 and \$3, depending on the service), and then there is a low per packet cost (a packet is 128 bytes, that is, 1 kilobit). As a result, i-mode is a major service of Internet email, online banking, ticket purchasing, game-playing, and peer-to-peer video messaging and gaming, with over 40 million users since 2003. There is a variety of Internet access plans available to mobile subscribers in the USA. For example, the Cingular GSM operator charges a fee of (\$0.01 per 1 kilobyte), with the cost decreasing with prepaid monthly plans.

1.3 The mobile Web stakeholders

The parties that will benefit from the growth of the mobile Web include, first, the end users, and then, the enterprises that will be able to provide new connectivity capabilities to their employees, the mobile network operators that will deploy the enabling networks,

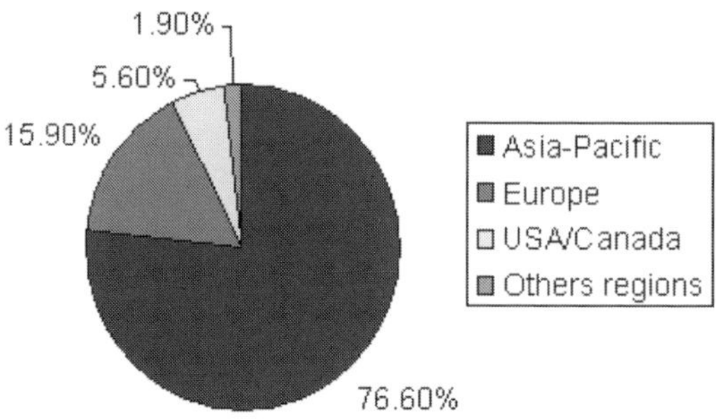

Figure 1.2 Mobile data subscribers by region.
(Source: © 2003 EMC. All rights reserved.)

the content providers that will reach their customers through the wireless Internet channel, and finally the mobile network and handset manufacturers that will build the support network elements and deliver enhanced handsets that will allow for an enjoyable end-user experience.

The mobile user

Future visions for the wireless Internet user have been articulated by software companies such as Sun and Microsoft, and adopted by mobile operators such as NTT DoCoMo. Significant growth is expected in specialized mobile services such as driving directions, traffic reports, tour guides, and any location-based services such as mobile shopping. In related user scenarios, the mobile terminal becomes the "network computer" and the "wallet PC". It is both an information device and a tool for enacting transactions based on the retrieved information. For example, users will use their mobile terminal to search and order products to be delivered at a nearby store, and, once at the store, they will use emoney to pay for the products with their mobile terminal.

As of September 2003 there were over 100 million wireless Internet users, with the majority in Japan and Korea, while fast growth rates were experienced in Europe. Figure 1.2 shows the distribution of mobile data users in 2003.

The enterprise

In the quest to enable employees, for example, sales staff, to be as productive when out of the office as when they are in the office, many enterprises support technologies and services that allow for "anywhere and anytime" connectivity to office information sources. This type of connectivity is made possible through wireless Internet solutions that leverage Web services [7] in the enterprise's domain. Web services provide a standards-based framework for integrating an enterprise's applications. They can supplant any existing middleware, enable programmers to interweave a company's applications as well as applications provided by external services, and reduce the reliance on proprietary solutions to implement application integration solutions.

The mobile network operator

Mobile carriers often have competed on subscription prices and offers of more minutes or free minutes during off-peak hours. The carriers differentiating themselves on service offerings will drive the future. In the voice world, many operators are now attempting to replicate Nextel's success with its unique Direct Connect voice service that establishes, at the touch of a button, a walkie-talkie like radio connection with one or more users for the exchange of short voice messages. This service simplifies connectivity for workgroups that often need to be in touch with each other, and can be used by groups of non-work related users, for example, a family on a vacation trip.

In the wireless Internet space, operators can differentiate themselves by the data services that they provide. Operators may choose to develop their own unique content, purchase specialized content from third party providers, or else provide a platform environment, referred to as a mobile ecosystem, for delivery of services offered by external parties. NTT DoCoMo, for example, chose the latter option and is providing billing services as well as a portal site for mobile service access. On the other hand, virtual network operators or operators of special purpose networks that operate a vertical market system may choose to have closer control of the content they provide to their subscribers.

In the "open" mobile ecosystem environment, the network operator can provide service management facilities such as service usage tracking and customer billing. Some research studies estimate that an operator's revenues from data services could be at least double those that are achieved in a "closed" model [5]. User ownership is perceived to be a key ARPU driver. To prevent erosion of value through service openness, the network operator can own the subscriber by managing the user's identity, the user's personal information (for example, content preferences), network presence, and user location.

NTT DoCoMo has pursued its goal to expand i-mode's "open" model in markets outside Japan. Through investments in non-Japanese operator shares, DoCoMo has formed alliances with operators to offer the i-mode service in markets that include Taiwan (KG Telecommunications), the Netherlands (KPN Mobile), the USA (AT&T Wireless), Spain (Telefonica Mobiles), and others, and as of July 2004 there were over three million i-mode users outside Japan [3]. DoCoMo sets up licensing agreements with mobile operators to provide them with patents, technologies and the know-how necessary to launch the i-mode service. DoCoMo has also played a major role in helping operators establish close ties with content providers and equipment vendors, which is essential in order to build a strong foundation for the i-mode service.

The content provider

A mobile user accesses business functionalities via Web services offered by content providers. A Web application may often combine multiple Web services, so that the application appears as one integrated business function. For example, a number of Web services used in a supply-chain architecture can be presented as a single application such as in the case of a product ordering system. Similarly, a restaurant reservation service can operate in conjunction with a restaurant review service, and a credit card processing system. Content providers could provide the Web services components or else take on their integration and offer the resulting comprehensive applications.

The mobile network and handset manufacturer

Networks that enable mobile terminals to access Web content require network elements with dedicated functionality to support the data transmissions, screen content generation, and contextual awareness of user situations, for example, their location. Corresponding support is needed in the handsets. Prior generations of mobile terminals provided small text-oriented displays with a few short lines of text that did not present attractive interfaces for information display via the installed micro-browsers. The newer mobile terminals have larger graphics-oriented screens, many with color, which enhance the wireless experience and entice users to access the Internet more often. Some operators come up with the specifications of the phones they wish to offer their subscribers. For example, NTT DoCoMo placed orders for phones with specific requirements with Japanese handset manufacturers. These phones include rich feature sets that support the goal of making the wireless Internet access experience effortless and enjoyable.

1.4 Mobile Web networks

The future of the mobile Internet will depend very largely on the ability of the wireless industry to deliver valuable services that are easy to use. Some of the major challenges for this success include the establishment of a standard user identification method, financial transaction standards for mobile commerce, and easy inclusion of third party service offerings in mobile networks. The ubiquitous availability of operating systems such as Windows and Mac OS, have facilitated the tremendous growth of application offerings on the desktop computer. A similar growth in mobile applications will occur as network operators open their networks to third party service providers.

1.4.1 The mobile network operating system

Mobile network operators could consider themselves as a component of the "mobile network operating system" where they provide services such as user location and user identification that are leveraged by the mobile applications. The other components of this mobile network operating system include the user mobile terminal and the application server that hosts the mobile applications, each providing its respective operating system services. For example, the mobile terminal provides a browser user interface, and the application server provides services for generating markup that is ultimately displayed by the mobile terminal's browser. There are additional service components as well, and these will be described in future chapters.

Growth of the mobile applications market will be assured when application developers will have straightforward access to all service components of the mobile network operating system: the user terminal capabilities, the network operator services, and the application server features. Operators will be able to differentiate themselves by the type of services that they provide to application developers. For example, if an operator has unique services that track the mobile user's context (e.g., whether the user is currently driving a vehicle), then the application provider will be able to better adapt an application's behavior to the user's changing context. This ability can directly affect the "ease of use" factor that determines whether users will adopt wireless Internet services.

The mobile network operating system will consist of a collection of the above services that are implemented as Web services and can be called by application developers. Web services were promoted as a set of shared protocols that enable disparate systems to talk to each other. These protocols include a Web Services Description Language (WSDL) that describes a service's interfaces, a Simple Object Access Protocol (SOAP) for transporting XML messages, and a Universal Description, Discovery, and Integration (UDDI) directory for storing information about Web service offerings. Web services can replace less flexible methods for information exchange, such as EDI (Electronic Data Interchange), that are used for the exchange of specific transaction data. With Web services multiple WSDL interfaces can be defined for accessing a service, and multiple clients can make use of the provided access methods.

1.4.2 Web services proposition

It is worthwhile revisiting the original motivation for Electronic Data Interchange (EDI) [8] to better understand how its Web services [7] successor enables building mobile network operating systems. EDI became a standard under the X12 American National Standards Institute (ANSI) committee. By December 2000, this standard defined over 300 transactions for the exchange of electronic documents, such as purchase orders, invoices, health care claims, or requests for proposals. Each EDI transaction consists of the transmission of a set of data segments, which are related data elements. For example, date and time are expressed in a separate data segment. With the definition of the electronic data forms exchanged between trading partners came an improvement in the overall business processes. The improvement was mainly due to the reduced intervention by humans. However, EDI is focused on the data forms being exchanged, and whenever these forms change, updates are required to the supporting software. Furthermore, the EDI transactions are not directly integrated with a trading partner's host applications. Separate software is needed to extract business data from an EDI transaction and make it available for use by other business applications.

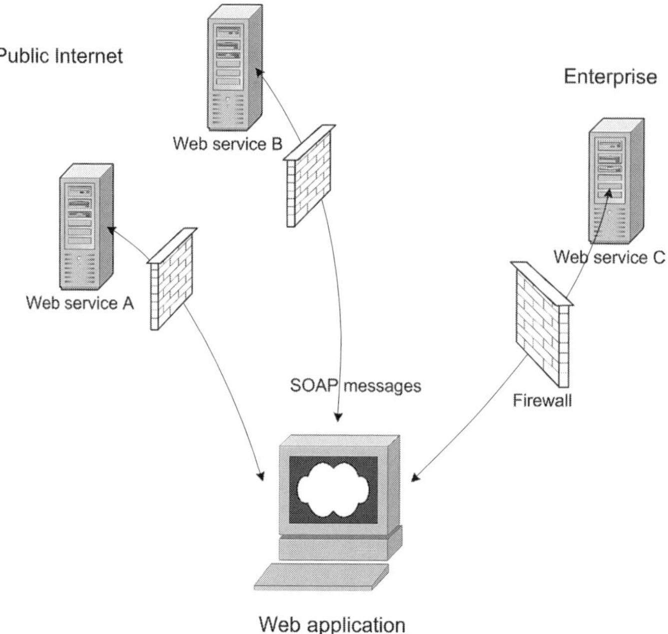

Figure 1.3 A Web application: integrating Web services with SOAP messages.

Unlike EDI, Web services provide a procedure call-like interface where services can be invoked in a way similar to the remote procedure call (RPC) of programming languages. Web services enable the building of software applications that execute on the Internet and use the same software paradigms that were successfully applied in the development of enterprise applications. For example, object-oriented development approaches provide for the reuse of software components, the objects that are the application's building blocks. A Web service is like an object that can be reused by multiple Web applications. A Web application is therefore considered to consist of collaborating Web services. To collaborate, these services rely on standard interfaces in the form of SOAP messages that convey requested operations and are transported on top of the standard HTTP protocol (see Figure 1.3). This interface is both a strength, as messages can then cross enterprise firewalls, and a weakness, as security needs to be enhanced to disallow malicious SOAP messages.

The reusability of Web services components and the relatively easy integration of Web services through standard interfaces are a very potent proposition for Application Service Providers (ASPs). ASPs were not very successful in the past as they had to spend large amounts of their resources on customizing their applications to specific customer needs and integrating them with pre-existing enterprise applications in the customer's premises. The license fees charged by ASPs covered only 20 percent of the application's deployment cost [9]. The ASPs did not realize that customization and integration could amount to as much as 80 percent of an application's implementation effort.

Some ASPs attempted to respond to these economic challenges by partnering with system integrators, and outsourcing services such as application hosting and application management. The advent of Web services-provided a new technology enabler that can significantly reduce the high costs of customization and application integration. Web services-based architectures encourage the development of specialized services. Such services can be easily added, removed, or replaced in a Web application so that customization is now concerned with how to glue together focused Web services through appropriate SOAP messages. Integrating a Web application with pre-existing enterprise applications is also answered by leveraging SOAP messaging.

General Motors, for example, adopted a Web services architecture in the implementation of its wireless OnStar car service [10]. From an automatic notification system of air bag deployment for dispatching help services to distressed drivers, OnStar evolved to provide driver data services such as step-by-step navigation, and online concierge services for event tickets, dining reservations, gift recommendations, etc. New services are built as Web services and can be integrated in OnStar more quickly and at a lower cost, without disrupting existing services. Some services, such as driver navigation, also leverage human advisors for the delivery of travel information via cellular calls.

1.5 Context-aware mobile services

The use of context in mobile applications has similar incentives to its use in fixed environments such as the desktop. These incentives include lowering the overhead of application use, and thereby improving the usability and adoption of applications. For example, in a desktop application a user may be presented with multiple tool bars and icons that offer too many services to master. Toolbars that take into account the user's role, responsibilities, and knowledge level, are much more effective from an end-user's perspective. In a mobile environment, the role of context as a service enabler is even more prominent. The user has a screen-constrained terminal, and is further limited by the interaction means available. For example, data may have to be entered using a phone's keypad.

A mobile application that is attuned to a user's context will leverage knowledge about *who* the user is, *what* the user is doing, *where* the user is, and *what* terminal the user is using. The application will be able to adapt itself to a user's intentions and thereby increase its effectiveness and acceptance as a useful solution. This condition-dependent service is also referred to as context-aware computing [11].

1.5.1 Personalized mobile user experience

Context-awareness has a major role to play in helping reach the goal of a personalized experience when the user interacts with mobile services. Forrester Research [12] has adapted the *Johari Window* [13] model of interpersonal processes to model the types of context in the interaction between user and application (see Figure 1.4). In this window,

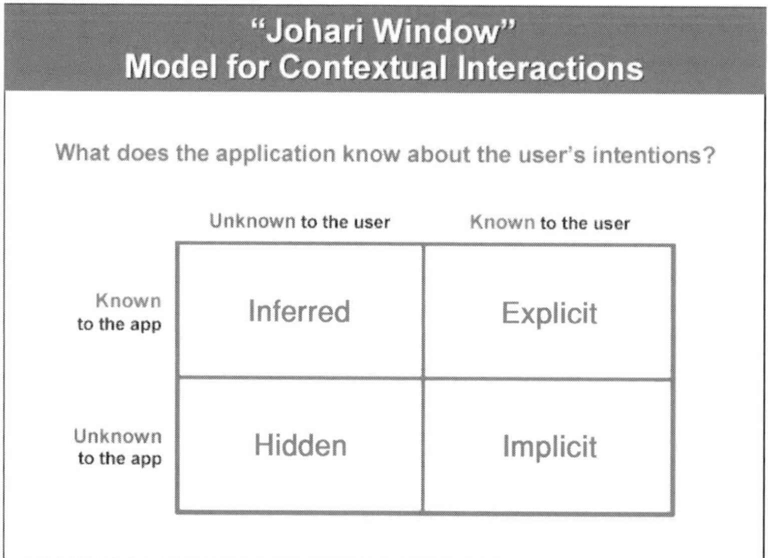

Figure 1.4 Context knowledge in user and application interactions.
(Source: © 2001 Forrester Research. All rights reserved.)

there are various states of context knowledge that are pertinent to both the user and the application. In the *Explicit* quadrant all available context is known to both sides, so little improvement can be achieved here. However, and this is the more prevalent case, either the user or the application are not fully cognizant of all context. In the *Inferred* quadrant, the application is knowledgeable about environment conditions, not known to the user, which can impact the application's results. For example, traffic conditions known to the application can affect the choice of suggested routes that users should take to reach their respective destinations. In the *Implicit* quadrant, the user has preferences that are not fully revealed to the application; however, these can be derived through analysis, for example, by checking the user's history of previous choices. Finally, in the *Hidden* quadrant, context needs to be guessed, and predefined rules can be used to best pick application choices. Time-based information, for example, can be used to decide when to send notifications; the fact that it is noon can trigger the sending of suggested places to eat lunch in the user's vicinity.

As the context quadrants of Figure 1.4 show, there is much room for creativity in the design of mobile services that want to leverage context knowledge. Mobile applications will be differentiated based on how well they are able to adapt their results to fit the user's current needs through knowledge of external environment conditions, through discovery of a user's implicit preferences, or through rules that propose best actions. Judicious use of context can help personalize the applications so that users receive the impression that they are dealing with a personal concierge rather than with a mass market service.

1.5.2 Elements of context

Context refers to elements such as:

- tasks that the user wishes to accomplish;
- absolute location;
- relative location (to landmarks);
- physical conditions of the environment (noise, brightness, etc.);
- movement relative to surroundings (for example, is the user in a driving car?);
- proximity to other users;
- User personal profile and habits;
- Context history (for conjecturing on the next context).

Some of these context elements, for example the tasks that the user wishes to accomplish, are specified by the mobile user. Others require sensing elements in the terminal or infrastructure, for example a GPS receiver in the mobile terminal can provide absolute location, while the infrastructure can detect relative location such as proximity to a park's attraction, or to a store in a shopping mall.

Context is typically partitioned into a number of categories. Chen and Kotz describe possible categorizations [14], and the categorization we use in this book partitions context elements between the user, network connectivity, and the environment:

1. User *static* context: specified in a user's profile and describes information interests and preferences.
2. User *dynamic* context: a user's location, current activity (walking, driving, etc.), and task (city touring, shopping, etc.).
3. Network connectivity context: network characteristics and mobile terminal capabilities.
4. Environmental context: time of day, noise, lighting, weather, etc.

Location is the most widely used context item in context-aware computing. Different methods can be used for tracking mobile users both outdoors and indoors. Outdoor methods include terminal-based systems such as GPS that can provide an accuracy of 10–20 meters, and network-based systems where cell sites use triangulation algorithms to determine a user's location. Indoor systems include infrared (IR) systems that send beacons of location data [15], and RF systems where the terminal measures the differential time of arrival between different radio signals, or measures the RF signal strength. Alternatively, a terminal can use connectivity-based schemes to determine a coarse-grain location, for example, proximity to a wireless local area network (WLAN) access point or to a Bluetooth access point.

Connectivity context refers to the connected network's identifier, available bandwidth, quality of service, cost per data packet, and mobile terminal features. The mobile client expects that a serving application will adapt to the network it connects to when it roams between networks. The format of the content sent to a user may change, for example, it could change from video in a WLAN environment, to text or audio in a wireless wide area network (WWAN) with smaller bandwidth capacity. This service adaptability is yet another enabler of what is referred to as *seamless mobility* where no explicit user actions are required to maintain an application's session when roaming between different networks.

Other types of context that can be sensed, and have been addressed in research projects (for example [16], [17], [18] and [19]), include time of day, vicinity to other people or objects, terminal orientation, light level, sound, temperature, and pressure. A user's current task, information interests, and communication preferences, are also considered to be context information. The user's current task could be determined by the application he or she has most recently invoked. On the other hand, a user's information interests are typically specified by the user manually filling a form that contains attributes listing possible information interests.

1.5.3 Context collection and storage

The process of detecting a user's context is highly dependent on the particular environment and the available sensors. As different sensors may be used in different environments, the applications that use context information should be developed in a manner independent from methods used to collect sensor data, interpret the data, and translate it to application-understandable formats (see, for example [20]). Context information generation from sensor data is also referred to as sensor fusion [21], and can be approached in two ways: centralized, where context is collected in one central store, or distributed, where context is maintained in a number of locations.

A centralized server can collect raw sensor information and provide interpreted context information via a standard API, or via events sent to applications that registered with the server. Alternatively, a directory, for example an X.500 directory service, accessed by the lightweight directory access protocol LDAP, can provide location and other context information. The downside of using directories is that they are not optimized for frequent updates, as in the case of context tracking, and require the interested applications to poll the directory to detect any context change.

In a distributed scheme, context is maintained in multiple locations including the client terminal. This puts more effort on applications, as they need to collect context information from a variety of locations. Intelligent agents can be used to gather context information for applications shielding the applications from the task of polling context stores or processing multiple context-related events.

The enablers of context include context engines that host context-aware applications. Forrester Research has defined a context engine as a platform that includes a variety of sensors that input context data, inference capabilities that draw upon context knowledge, and adaptive interfaces to users or machines [22]. A context engine can host context-aware applications, and is therefore seen as a specialization of application servers.

1.6 Standards bodies

The standards referenced in this book attest to the fact that there are many standards bodies that contribute to the wireless industry, its communication protocols, languages and architectures for mobile services development. The standards bodies which are referenced in this book, and their stated goals, include:

1. 3GPP
   The 3rd Generation Partnership Project (3GPP) is a collaboration agreement that was established in December 1998. The collaboration agreement brings together a number of telecommunications standards bodies which are known as "organizational partners". The original scope of 3GPP was to produce globally applicable technical specifications and technical reports for a 3rd Generation Mobile System based on evolved GSM core networks and the radio access technologies that they support (i.e., Universal Terrestrial Radio Access (UTRA) both Frequency Division Duplex (FDD) and Time Division Duplex (TDD) modes). The scope was subsequently amended to include the maintenance and development of the Global System for Mobile communication (GSM) technical specifications and technical reports including evolved radio access technologies (e.g., General Packet Radio Service (GPRS) and Enhanced Data rates for GSM Evolution (EDGE)). The 3GPP produces technical specifications and reports and can be accessed at http://www.3gpp.org.
2. 3GPP2
   The Third Generation Partnership Project 2 (3GPP2) is "a collaborative third generation (3G) telecommunications specifications-setting project comprising North American and Asian interests developing global specifications for ANSI/TIA/EIA-41 Cellular Radiotelecommunication Intersystem Operations network evolution to 3G and global specifications for the radio transmission technologies (RTTs) supported by ANSI/TIA/EIA-41." 3GPP2 was born out of the International Telecommunication Union's (ITU) International Mobile Telecommunications "IMT-2000" initiative, covering high speed, broadband, and Internet Protocol (IP)-based mobile systems featuring network-to-network interconnection, feature/service transparency, global roaming, and seamless services independent of location. The 3GPP2 produces specifications and can be accessed at http://www.3gpp2.org/.
3. ANSI
   The American National Standards Institute (ANSI) is a private, non-profit organization, founded in 1918, that administers and coordinates the US voluntary standardization and conformity assessment system. The Institute's mission is to enhance both the global competitiveness of US business and the US quality of life by promoting and facilitating voluntary consensus standards and conformity assessment systems, and safeguarding their integrity. ANSI can be accessed at http://www.ansi.org/.

4. FIPA

   The Foundation for Intelligent Physical Agents (FIPA) was formed in 1996 to produce software standards for heterogeneous and interacting agents and agent-based systems. FIPA produces specifications and can be accessed at http://www.fipa.org/.
5. GSMA

   The Global System for Mobile Communications Association (GSMA) is a global trade association serving the world's GSM mobile operator member community by promoting, protecting and enhancing their interests and investments. The GSMA produces requirements and can be accessed at http://www.gsmworld.com/.
6. IETF

   The Internet Engineering Task Force (IETF) is a large open international community of network designers, operators, vendors, and researchers concerned with the evolution of the Internet architecture and the smooth operation of the Internet. The IETF produces requests for comments (RFCs) and can be accessed at http://www.ietf.org.
7. ISO

   The International Organization for Standardization (ISO) is a network of the national standards institutes from 148 countries, on the basis of one member per country, with a central secretariat in Geneva, Switzerland, that coordinates the system. ISO began operations in 1947 with the objective of facilitating the international coordination and unification of industrial standards. ISO generates standards and can be accessed at http://www.iso.ch.
8. ITU-T

   The ITU Telecommunication Standardization sector (ITU-T) is one of the three sectors of the International Telecommunication Union. ITU-T's mission is to ensure an efficient and on-time production of high quality standards (recommendations) covering all fields of telecommunications. The ITU-T produces recommendations and can be accessed at http://www.itu.int/ITU-T/.
9. Liberty Alliance

   The Liberty Alliance Project was formed in September 2001 to develop open standards for federated network identity management and identity-based services. Its goals are to ensure interoperability, support privacy, and promote adoption of its specifications, guidelines and best practices. The Liberty Alliance generates specifications and can be accessed at http://www.projectliberty.org/.
10. OASIS

    The mission of the Organization for the Advancement of Structured Information Standards (OASIS) is to drive the development, convergence, and adoption of structured information standards in the areas of ebusiness, web services, etc. OASIS produces specifications through the work of its technical committees and can be accessed at http://www.oasis-open.org/.
11. OMA

    The mission of the Open Mobile Alliance (OMA) is to facilitate global user adoption of mobile data services by specifying market driven mobile service enablers that ensure service interoperability across devices, geographies, service providers, operators, and networks, while allowing businesses to compete through innovation and differentiation. The OMA produces specifications and can be accessed at http://www.openmobilealliance.org/.

12. W3C
    The World Wide Web Consortium (W3C) develops interoperable technologies (specifications, guidelines, software, and tools) to lead the Web to its full potential. W3C is a forum for information, commerce, communication, and collective understanding. The W3C produces recommendations and can be accessed at http://www.w3c.org.

1.7 XML and XML Schema

Most of the code examples presented in this book are based on the Extensible Markup Language (XML), a document-structuring meta-language with which one can define a set of tags that provide structure and meaning when added to a document. XML 1.0 was released by the W3C in February 1998 [23], as a simplified successor to the Standard Generalized Markup Language (SGML) specified by the International Organization for Standardization (ISO) in 1986 [24]. A simplified version of SGML is Hypertext Markup Language (HTML) [25], in use since 1990 for specifying information formatting for Web documents. Although HTML is simple to use, it is limited by the fact that it has a fixed set of tags, and the tags do not convey meaningful information for content search purposes. As the first objective of the Web was to provide human readable documents that could be easily shared, HTML was a perfect fit for meeting this goal. With the next W3C objective to create a semantic Web, HTML was no longer a good choice for document markup. In the semantic Web, machines inter-communicate to search and exchange documents. This latter goal can be met only with a markup that attaches meaning to documents and their sub-components.

HTML is considered a specific XML vocabulary, and other vocabularies can be defined in the same way. An XML entity is referred to as a document and the document contains XML elements. Each element has a start tag and an end tag, and may contain character data, child elements, or both. Alternatively, an element can be defined by a single empty tag. For example, the XML entity in Listing 1.1 defines a database of hotels.

```
<?xml version="1.0"?>
<hotelData updateDate="2004-02-05">
<!-- Last update done by Larry -->
  <hotels>
    <hotel name="Harbor Lite">
      <lodgingType>Motel</lodgingType>
```

```
[7]          <units>64</units>
[8]          <twoPersonPrice>70.00</twoPersonPrice>
[9]          <pool>true</pool>
[10]         <comment>Check if year-round pool</comment>
[11]       </hotel>
[12]       <hotel name="The Grey Whale">
[13]         <lodgingType>BedAndBreakfast</lodgingType>
[14]         <units>15</units>
[15]         <onePersonPrice>39.00</onePersonPrice>
[16]         <twoPersonPrice>60.00</twoPersonPrice>
[17]         <minimumStay>2nights</minimumStay>
[18]       </hotel>
[19]     </hotels>
[20] </hotelData>
```

Listing 1.1 XML hotel database

Some elements in Listing 1.1 include attributes specified by a name followed by a value enclosed in quotes. Attributes provide further detail about elements, for example, the attribute *updateDate* in the *<hotelData>* element specifies the date when the database was last updated. An attribute declaration in an XML schema indicates whether a declared attribute is required in the containing element or may be absent.

The XML code in Listing 1.1 also includes a processing instruction in the first line that indicates that the document is based on XML version 1.0. As with comments in programming languages, comments can be included in XML; they begin with *<!--* and end with --> as shown in line 3. Alternatively, as is done in line 10, the XML author can define a *<comment>* element to contain any notes to be embedded in the document. Another, often used, concept in XML is the notion of namespace. A namespace definition associates a URL with a given name prefix. All names with the same prefix are defined at the specified URL. This helps avoid name conflicts when similar names from different sources are used in the same XML document. For example, Listing 1.2 includes in the first line a declaration of a namespace prefix *xsd*: which is associated with the URL of the XML Schema definition.

The structure of XML code can be specified with the XML Schema definition language which is represented in XML as well ([26], [27]). With XML Schema, the XML author can express syntactic, structural, and value constraints applicable to the document elements. For example, XML Schema can specify what elements are allowed in a document, whether they are required to appear and if there is an upper bound on the number of an element's occurrences. Element content type, for example, string, Boolean, decimal, or float, is specified by the schema. XML Schema also specifies the attributes of elements, whether they are required or optional, and their content type.

The schema of the above hotel database is shown in Listing 1.2. The *<hotelData>* element is of type *xsd:complexType*, meaning that it contains other elements listed in the corresponding *<xsd:sequence>* element. This element also contains a required attribute that specifies a date and is declared with the *<xsd:attribute>* element. One of the elements contained in the *<hotelData>* element, *<hotels>*, is itself a complex element as it contains the list of *<hotel>* elements. There can be zero or an unbounded number of *<hotel>* elements as specified by the *minOccurs* and *maxOccurs* attributes in the *hotel* declaration element. Each *<hotel>* element has a required attribute *name* declared in line 35 that gives the hotel name.

The content of XML elements can be restricted. For example, possible values can be restricted to an enumerated list of values as in the declaration of the *<minimumStay>* element in lines 27–31, where the hotel requires a minimum stay that can be two, three, or four nights.

```
[1]  <xsd:schema xmlns:xsd=“http://www.w3.org/2001/XMLSchema”>
[2]
[3]  <xsd:element name=“hotelData” type=“hotelDataType”/>
[4]  <xsd:element name=“comment” type=“xsd:string”/>
[5]
[6]  <xsd:complexType name=“hotelDataType”>
[7]  <xsd:sequence>
[8]    <xsd:element ref=“comment” minOccurs=“0”/>
[9]    <xsd:element name=“hotels” type=“Hotels”/>
[10] </xsd:sequence>
[11] <xsd:attribute name=“updateDate” type=“xsd:date”
       use=“required”/>
[12] </xsd:complexType>
[13]
[14] <xsd:complexType name=“Hotels”>
[15] <xsd:sequence>
[16]   <xsd:element name=“hotel” minOccurs=“0”
         maxOccurs=“unbounded”>
[17]   <xsd:complexType>
[18]   <xsd:sequence>
[19]     <xsd:element name=“lodgingType” type=“xsd:string”/>
[20]     <xsd:element name=“units” type=“xsd:positiveInteger”/>
[21]     <xsd:element name=“onePersonPrice”
           type=“xsd:decimal”/>
[22]     <xsd:element name=“twoPersonPrice”
           type=“xsd:decimal”/>
```

```
      <xsd:element name="pool" type="xsd:boolean"
        minOccurs="0"/>
      <xsd:element ref="comment" minOccurs="0"/>
      <xsd:element name="minimumStay" minOccurs="0">
        <xsd:simpleType>
          <xsd:restriction base="xsd:string">
            <xsd:enumeration value="2nights"/>
            <xsd:enumeration value="3nights"/>
            <xsd:enumeration value="4nights"/>
          </xsd:restriction>
        </xsd:simpleType>
      </xsd:element>
    </xsd:sequence>
    <xsd:attribute name="name" type="xsd:string"
      use="required"/>
    </xsd:complexType>
    </xsd:element>
  </xsd:sequence>
  </xsd:complexType>

  </xsd:schema>
```

Listing 1.2 XML Schema of hotel database

The content of an XML document instance can be validated against a corresponding schema definition by submitting both documents to an XML validation program. Prior to XML Schema, XML authors used Document Type Definitions (DTDs) to express the structure of XML documents. DTD support is provided in most XML parser programs. While the language used for DTD statements is concise, it is not XML-based. The language addresses mainly the structure of XML documents, and has no support for defining varied content types that can be included in elements and attributes. The specification of the DTD language is included in the W3C XML specification. The DTD that corresponds to the XML Schema of the above hotel database is shown in Listing 1.3.

```
<!ELEMENT comment (#PCDATA)>

<!ELEMENT hotel (lodgingType, units, onePersonPrice,
   twoPersonPrice, pool?, comment?, minimumStay?)>
<!ATTLIST hotel name CDATA #REQUIRED>

<!ELEMENT hotelData (hotels)>
```

```
[7]  <!ATTLIST hotelData updateDate NMTOKEN #REQUIRED>
[8]
[9]  <!ELEMENT hotels (hotel*)>
[10]
[11] <!ELEMENT lodgingType (#PCDATA)>
[12]
[13] <!ELEMENT minimumStay (#PCDATA)>
[14]
[15] <!ELEMENT onePersonPrice (#PCDATA)>
[16]
[17] <!ELEMENT pool (#PCDATA)>
[18]
[19] <!ELEMENT twoPersonPrice (#PCDATA)>
[20]
[21] <!ELEMENT units (#PCDATA)>
```

Listing 1.3 DTD of hotel database

The DTD code in Listing 1.3 declares each XML element with the *<!ELEMENT>* tag. The *<hotel>* XML element in line 3 is declared with a list of the required and optional elements that it can include; optional elements use the character *"?"* as a name terminator. The *<hotel>* element also includes a required attribute of type character data (CDATA). The following declaration for *<hotelData>* also includes an attribute, this time of type name token (NMTOKEN).

The *<hotels>* element can contain zero or more *<hotel>* elements as specified by the character "*", and each of the following element declarations define elements that can contain only strings that are referred to as parsed character data (PCDATA).

REFERENCES AND FURTHER READING

[1] Surfline, http://www.surfline.com/home/index.cfm.

[2] Cellular-news, *Openwave claims 400 million WAP browser sales*. (http://www.cellular-news.com/story/9404.shtml, Jul. 30, 2003).

[3] Cellular-news, *i-mode users surge upwards*. (http://www.cellular-news.com/story/11439.shtml, Jul. 12, 2004).

[4] Google, *Google Wireless Services*. (http://www.google.com/wireless/, 2004).

[5] E. Steels, *Creating and Sharing Value in the Mobile Ecosystem*, Openwave. http://www.openwave.com, Sep. 2001.

[6] T. Natsuno, *i-mode Strategy* (Chichester, England: John Wiley and Sons Ltd, 2003).

[7] F. Curbera *et al.*, Unraveling the Web Services web. *IEEE Internet Computing* (Mar/Apr. 2002).

[8] EDI, http://www.x12.org.

[9] J. Hagel III, *Out of the Box* (Boston, MA: Harvard Business School Press, 2002).

[10] OnStar Corporation, http://www.onstar.com.

[11] B. Schilit *et al., Context-aware computing applications*. IEEE workshop on mobile computing systems and applications, Dec. 1994.

[12] C. Zetie, The emerging context-aware software market: How applications will get smarter. *Forrester Research* (Nov. 5, 2001).

[13] J. Luft, *Of Human Interaction* (Palo Alto, CA: National Press, 1969).

[14] G. Chen and D. Kotz, *A Survey of Context-Aware Mobile Computing Research*, Technical report TR2000-381. Dept. of Computer Science, Dartmouth College (2000).

[15] A. Pashtan *et al.*, Adapting content for wireless Web Services. *IEEE Internet Computing* (Sep. 2003).

[16] G. Abowd *et al.*, Cyberguide: A mobile context-aware tour guide. *Wireless Networks*, **3** (1997), 421–33.

[17] K. Cheverst *et al.*, Experiences of developing and deploying a context-aware tourist guide: The Guide Project. *ACM MOBICOM* (2000).

[18] N. Davies *et al.*, Using and determining location in a context-sensitive tour guide. *IEEE Computer* (Aug. 2001), 35–41.

[19] A. Hinze and A. Voisard, *Location and Time-based Information Delivery in Tourism*. Proc. of the Intl Conf. on Spatio-temporal Databases (SSTD3), Lecture Notes in Computer Science N. 2750. (Springer Verlag, 2003).

[20] A. K. Dey *et al.*, A conceptual framework and a toolkit for supporting the rapid prototyping of context-aware applications. *Human–Computing Interaction (HCI) Journal*, **16**:2–4 (2001), 97–166.

[21] H. Wu *et al.*, Sensor fusion for context understanding. *IEEE Instrumentation and Measurement Technology Conf.* (May 2002).

[22] C. Zetie, Market overview: The emerging context-aware software market. *Forrester Research* (Nov. 7, 2001).

[23] T. Bray *et al., Extensible Markup Language (XML) 1.0*. W3C Recommendation (Feb. 4, 2004), http://www.w3.org/TR/2004/REC-xml-20040204.

[24] International Organization for Standardization, *Standard Generalized Markup Language (SGML)*. ISO 8879 (1986).

[25] D. Raggett *et al., HTML 4.01 Specification*. W3C Recommendation (Dec. 24, 1999), http://www.w3.org/TR/html401/.

[26] H. S. Thompson *et al., XML Schema Part 1: Structures*. W3C Recommendation (May 2, 2001), http://www.w3.org/TR/2001/REC-xmlschema-1-20010502/.

[27] P. V. Biron and A. Malhotra, *XML Schema Part 2: Datatypes*. W3C Recommendation (May 2, 2001), http://www.w3.org/TR/2001/REC-xmlschema-2-20010502/.

T. Berners-Lee and D. Connolly, *HTML 2.0*, RFC 1866. IETF (Nov. 1995).

M. van Bekkum *et al.*, A user-centric design of a personal service environment. *3rd WWRF Proc.* (Sep. 2001).

P. Brezillon, Context in problem solving: A survey. *The Knowledge Engineering Review*, **14**:1 (1999).

G. Chen and D. Kotz, *Context Aggregation and Dissemination in Ubiquitous Computing Systems*, Technical report TR2002-420. Dept. of Computer Science, Dartmouth College (2002).

N. Davies *et al.*, The rational for infrastructure support for adaptive and context-aware applications: A position paper. *NSF workshop IMWS 2001*, Oct. 15, 2001. (Spinger Verlag, 2002).

Digital Thinking Network, *Scenarios for the Future of the Mobile Internet in Europe in 2007*. (http://www.dtn.net, Sep. 2002).

A. Dix *et al.*, Exploiting space and location as a design framework for interactive mobile systems. *ACM Trans. on Computer–Human Interaction*, **7**:3 (Sep. 2000), 285–321.

EC 5th. Framework Program, *M-ToGuide*. (2003), http://dbs.cordis.lu/fep-cgi/srchidadb?ACTION=D&CALLER=PROJ_IST&QF_EP_RPG=IST-2001-36004.

B. König-Ries *et al.*, Developing an infrastructure for mobile and wireless systems. *NSF workshop IMWS 2001*, Oct. 15, 2001. (Springer Verlag, 2002).

W. Ma *et al.*, A framework for adaptive content delivery in heterogeneous network environments. *Multimedia computing and networking 2000* (Jan. 2000).

A. Pashtan and J. Yanosy, An adaptable services framework. *5th. WWRF Proc.*, Mar. 2002.

T. Pham *et al.*, Exploiting location-based composite devices to support and facilitate situated ubiquitous computing. *HUC 2000, Lecture notes on Computer Science 1927* (2000), 143–56.

S. Ulfelder, GM gears up with collaboration based on Web services. *Network World* (http://www.nwfusion.com/research/2003/0526gm.html, May 26, 2003).

T. Winograd, Architectures for context. *HI Journal* (2001).

2 Wireless system architecture

This chapter introduces the wireless system architecture specified by the WAP Forum (now the Open Mobile Alliance – OMA). Protocol stacks and network components, including gateways, are described. The WAP Forum started its work in 1997, and its latest release is referred to as WAP 2.0. We proceed with a description of the i-mode wireless architecture of Japan's NTT DoCoMo operator, geared to support connectivity to the wired Internet. We show how both the WAP and i-mode architectures have converged to a wireless Internet architecture that leverages the same protocols used in the wired Internet domain, and where mobile terminals are regular Internet clients. Finally, we show how both WAP 2.0 and i-mode are realized over GSM's GPRS data network.

2.1 WAP 1.x architecture

The Wireless Application Protocol (WAP) Forum industry group was started in 1997 with the goal to specify a standard for presentation and delivery of wireless information and telephony services on mobile terminals [1]. The WAP Forum chose to align its standards closely with the Internet and the Web, and leverage their technologies such as communication protocols (for example, TCP/IP, HTTP, SSL) and presentation languages (for example, XHTML). Other major standard goals were to maintain air interface independence and device independence. These latter two goals would enable applications developed according to the WAP Forum standard to be portable across networks and devices.

2.1.1 The WAP 1.x protocol

Bandwidth limitations and noisy environments, when compared to the wired Internet, characterize wireless Internet access. In a noisy environment, interference will corrupt packets, and these packets will need to be retransmitted by the sender, which leads to high round-trip delays (latency). To overcome these issues, the WAP Forum promoted the concept of a wireless access protocol dedicated to Internet transmissions [2]. This

WAP protocol, between the mobile terminal and the wireless network infrastructure, was designed to be wireless bearer independent, and consists of multiple layers starting at the Transport layer through the Application layer. The WAP protocol layers from the bottom up provide support for:

1. Transport (Wireless Datagram Protocol – WDP) which is the interface between the wireless bearer and the WAP layers. WDP is a general datagram service. Multiple bearer networks can be supported including SMS, GSM, and CDMA. The WDP layer includes an Adaptation layer that is different for each bearer network.
2. Security (Wireless Transport Layer Security – WTLS) which provides data integrity, privacy, authentication, and denial-of-service protection, as specified in the IETF's Transport Layer Security (TLS).
3. Transaction (Wireless Transaction Protocol – WTP) which supports the push messages (guaranteed and not guaranteed) delivered from the network to the mobile terminal.
4. Session (Wireless Session Protocol – WSP). This is a binary-encoded version of HTTP 1.1, with some added functions and reduced latency capabilities for wireless support.
5. Application layer (WAP Application Environment – WAE). This layer includes the content specification and presentation languages, Wireless Markup Language (WML) and WMLScript, as well as a Wireless Telephony API (WTA).

All transmitted content between the mobile terminal and the network is compressed into a binary-encoded format. Graphic content is also in a WAP-specific binary form (Wireless Bit Map – WBMP). The purpose of the compression and the special graphic encoding is to improve transmission performance over the WAP Forum's over-the-air protocol. However, with a new over-the-air protocol that interfaces to the legacy wired Internet protocols, a WAP gateway is required to do the corresponding protocol conversions between the wireless and wired domain. The WAP 1.x network architecture is shown in Figure 2.1, and the respective protocol layers at the mobile terminal, gateway, and content server, are shown in Figure 2.2.

2.1.2 The WAP 1.x gateway

The WAP gateway functions include protocol conversions between wireless and wired domains, Internet data conversions and encoding/decoding between the wireless and wired domains, and miscellaneous control functions.

Protocol conversions

For any request or response, the gateway translates the message between WSP and HTTP formats. For example, a WSP request and associated request headers are sent by the user mobile terminal device to the gateway in a compact binary tokenized form that is translated into text encoded HTTP messages. In addition, the gateway may do

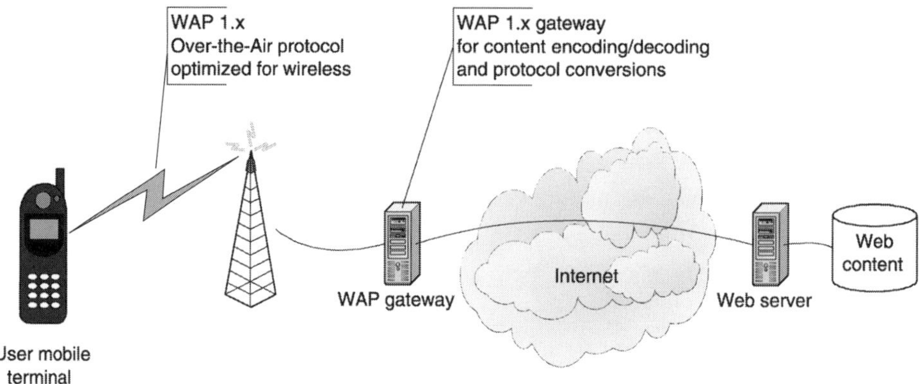

Figure 2.1 WAP 1.x architecture.

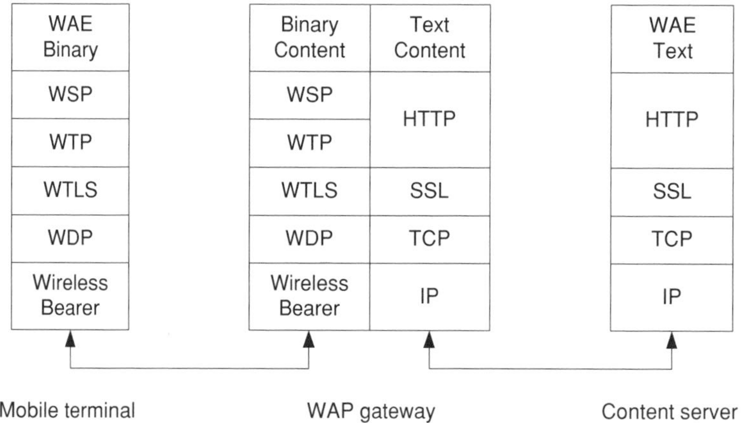

Figure 2.2 WAP 1.x protocol stacks.

data encodings between the wired and wireless domains to minimize the amount of data transmitted over the air.

Data conversions and encoding/decoding

Data conversion is an optional feature of WAP gateways. Wired Internet data in HTML form could be converted to WML suitable for display on the user's mobile terminal device. After this conversion, the WML data is encoded, through a tokenization process, into a compact binary form that can be efficiently transmitted over the air. Any WMLScript will be compiled too into an efficient bytecode representation that is interpreted on the client side. As it is difficult to formulate generic data conversion rules, Internet sites that want to provide content to mobile users in the WAP 1.x architecture often have data available in both HTML and WML formats.

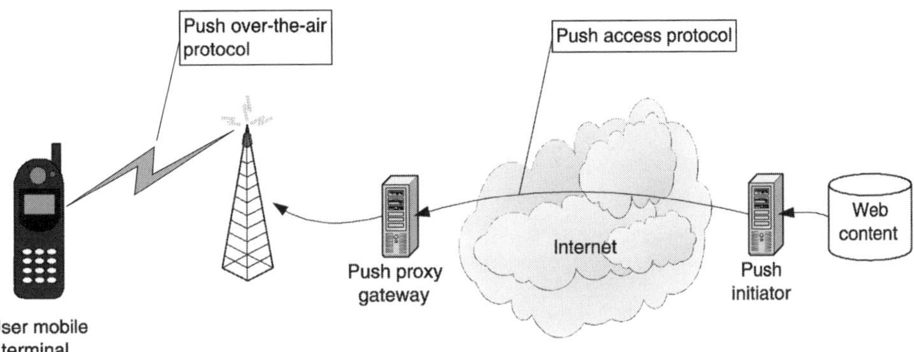

Figure 2.3 WAP 1.x push framework.

Miscellaneous control functions

The WAP gateway's control functions include terminal access checking, security, domain name resolution, data caching, service provisioning, and service utilization tracking. As the main function of the gateway is to provide efficient transportation of Internet content over the air, it is most likely to be owned by the network operator who wants to maximize his revenue from the use of the available radio frequencies. However, the access check and security control functions of a WAP gateway are functions that are usually dependent on the service requested by the user, and therefore could be controlled by the content provider. Access check would include the validation of the user's mobile terminal ID and the terminal IP address.

A gateway that implements security would use WTLS between the user mobile terminal and the gateway, and SSL or TLS between the gateway and the origin content server. As the gateway serves as an intermediary between these two security schemes, there may be cause for concern if applications require end-to-end encryption. Data will be temporarily unencrypted when the gateway converts between the two schemes.

2.1.3 WAP 1.x push

A unique capability introduced in the WAP 1.x architecture is the ability to *push* unsolicited content to wireless mobile terminals [3]. This enables applications to alert the user of time and location sensitive information, for example, a sale occurring in a store in close proximity, or seating availability in a nearby restaurant. This kind of information delivery is not available in standard HTTP where GET requests are used to *pull* information from a Web server. The WAP 1.x push framework is shown in Figure 2.3.

The Push Initiator submits content push submissions to the Push Proxy Gateway (PPG) using the Push Access Protocol (PAP) [4]. The PAP message is carried in the body of an HTTP POST message. Before sending the content, the Push Initiator may query the PPG for client capabilities and preferences to format appropriately the content for the client's display means. The PPG attempts to find the correct destination mobile terminal and delivers the push submission to that client using the Push Over-the-Air protocol [5]. To establish the connection between the PPG and the mobile terminal, the PPG would typically send the mobile terminal an SMS message with a Session Initiation Request (SIR). The mobile terminal would then respond by activating the appropriate bearer and contacting the PPG. As in the case of the WAP gateway, the PPG may binary-encode the received submission prior to its delivery to the mobile terminal.

In all content push submission messages, the Universal Resource Identifier (URI) of the content is sent to the mobile terminal. There are two major push submission messages. The *Service Loading* message [6] is interpreted by the browser to automatically retrieve and display the content pointed at by the message URI. The other message is a *Service Indication* [7]. In this case, the browser provides an indication to the user that content is available for display. The indication could take the form of a flashing icon or a short message. The user has the option of either loading the content immediately by issuing a corresponding GET request, or else postponing the handling of the *Service Indication* to a later point in time. There is yet another message related to content push, the *Cache Operation*, however, this message is a means to invalidate content in the mobile terminal cache, and does not deliver actual content.

2.2 WAP 2.0 architecture

The WAP Forum's WAP 2.0 specification was published in 2001 [8]. In WAP 2.0, new higher-speed wireless bearers that need to be supported include GSM's General Packet Radio Service (GPRS) and 3rd Generation (3G) cellular. This version of WAP represents further convergence to the Internet standards.

2.2.1 The WAP 2.0 protocol

As wireless communication migrates to 3G networks that exhibit much higher bandwidths, there is less need for the special purpose WAP 1.x protocols. The next generation WAP, WAP 2.0, has defined protocols that are more aligned with the wired Internet TCP and HTTP protocols. These protocols are the wireless-profiled TCP [9] and the wireless-profiled HTTP [10]. The WAP 2.0 network architecture is shown in Figure 2.4, and the respective protocol layers at the mobile terminal, gateway, and content server, are shown in Figure 2.5.

The wireless-profiled TCP (WP-TCP) is a connection-oriented protocol fully interoperable with TCP, whereas the WDP transport in WAP 1.x is a datagram-based protocol that requires conversion for communicating with TCP. WP-TCP supports both a split-mode and a regular TCP connection. In split-mode, the WAP gateway acts as an

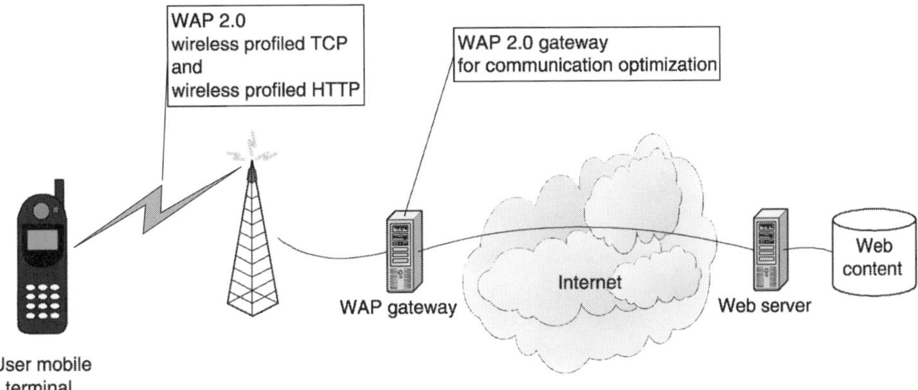

Figure 2.4 WAP 2.0 architecture.

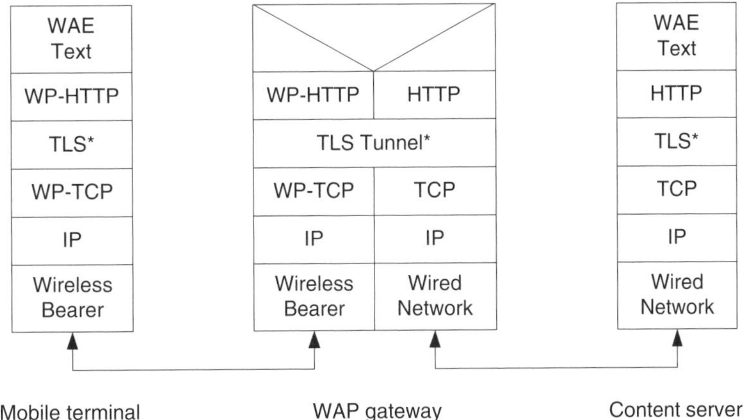

Figure 2.5 WAP 2.0 protocol stacks (split mode).

intermediary that interfaces between an optimized TCP for over-the-air transmissions and a regular TCP connection to the content server. The optimization includes adjusting TCP parameters for a better fit with air interface characteristics. For packet flow control, TCP uses a buffer, referred to as window, where the number of unacknowledged bytes sent is limited by the *window size* parameter. For example, WP-TCP implementations should support large window sizes so that TCP congestion control algorithms will be able to better find a congestion window size that is appropriate for the air interface.

The wireless-profiled HTTP (WP-HTTP) in the mobile terminal enables it to act as an HTTP client for requesting Web-based content and as an HTTP server to support *push* submissions from the network. These latter push submissions are implemented as HTTP POST messages. The mobile terminal can therefore issue GET and POST requests, and can handle POST requests sent to it. To support secure connections to

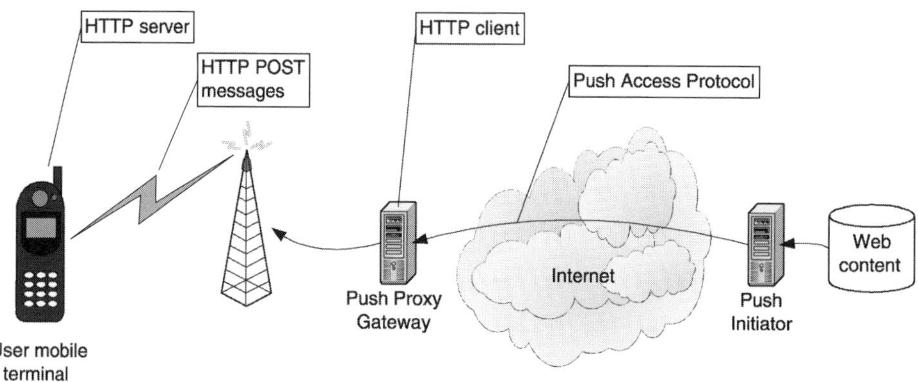

Figure 2.6 WAP 2.0 push framework.

content servers, the mobile terminal can issue HTTP CONNECT messages to establish a TLS tunnel to the content server via the WAP gateway as shown in Figure 2.5.

2.2.2 The WAP 2.0 gateway

In WAP 2.0 the user's mobile terminal issues regular HTTP requests, and the data displayed on the device is the same data available on the wired Internet, that is, HTML or XHTML text. Therefore, the WAP 1.x gateway protocol conversions and the WML binary encoding/decoding functions are no longer required. The wireless link supports wireless-profiled (WP) versions of TCP and HTTP ([9], [10]) that interoperate with the wired versions of these protocols, and the gateway hosts both versions of these protocols. WP-HTTP optimizations implemented by the gateway include support for compression of message bodies destined to the terminal for improved wireless link utilization.

In addition, the gateway supports the establishment of a secure tunnel between a mobile terminal and a content server when a secure connection is required. The TLS protocol is then used all the way between terminal and server, and the TLS packets are tunneled through the gateway.

2.2.3 WAP 2.0 push

Push submissions are implemented in WAP 2.0 as HTTP POST messages, and this implies that the PPG acts as an HTTP client and the receiving mobile terminal acts as an HTTP server, as shown in Figure 2.6. The sending of POST messages requires that a TCP connection be established first. Connection establishment can be initiated by the PPG, if the PPG knows the terminal's IP address and the wireless bearer is active or can be activated by the PPG. Otherwise, as in WAP 1.x, the PPG sends to the mobile

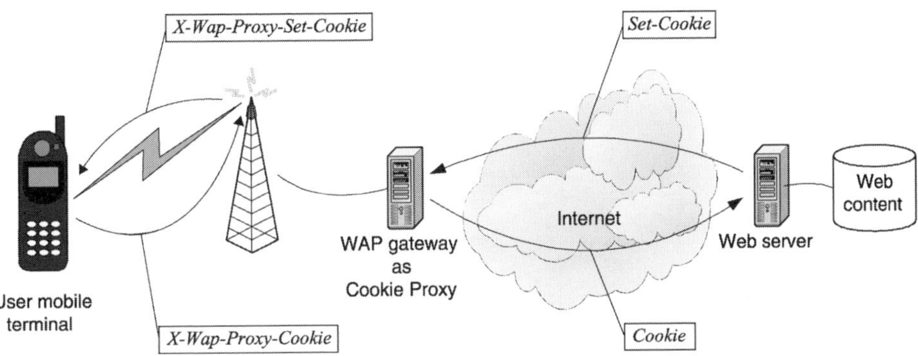

Figure 2.7 WAP 2.0 cookie proxy function.

terminal a Session Initiation Request, and the mobile terminal then establishes an active TCP connection towards the PPG.

2.2.4 WAP cookies

Since HTTP is a stateless protocol, the WAP Forum embraced the IETF concept of cookies for managing the state of a mobile terminal HTTP session. HTTP state management enables a content server to request that a small unit of state (a "cookie") is stored in the client, and included in subsequent requests to the origin server. As in the IETF defined protocol [11], the WAP Forum specified that the client's browser is responsible for cookie management and handling content server requests for cookie control [12]. Such requests can, for example, define the cookie's lifetime, and when the cookie should be included in client issued requests. Cookie related requests are conveyed in HTTP headers. The *Set-Cookie* header allows the content server to request a cookie to be stored for a certain predefined period of time. The *Cookie* header is used by the client's browser to deliver the cookie to the content server in subsequent requests, depending on whether the client's request URL contains a URL prefix specified in a previously received *Set-Cookie* header. The WAP specification requires that WAP browsers be able to store at least four cookies of 125 bytes each.

The WAP HTTP state management extended the network architecture with support for mobile terminals that have limited memory resources for cookie storage by defining the concept of a *cookie proxy* (see Figure 2.7). The cookie proxy, when present, is responsible for managing and storing cookies on behalf of the client, and modifies HTTP requests and responses to and from the client to implement this function. WAP specific HTTP headers were defined so that the client browser can enable or disable the storage of cookies in the proxy. The cookie proxy could be implemented in the WAP gateway for those clients that do not support local cookie storage. Additional benefits of relying on a proxy are that less information needs to be transmitted over the air as

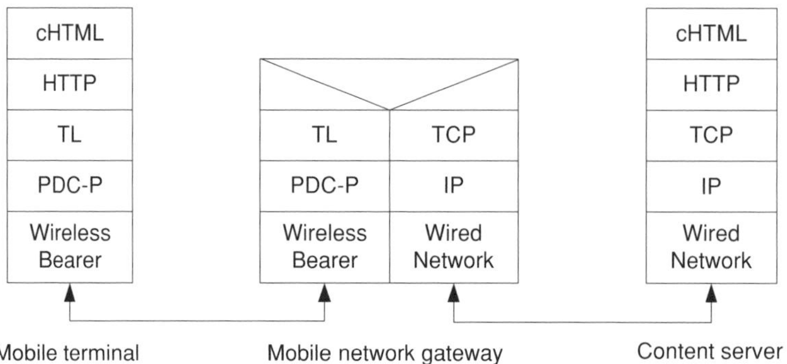

Figure 2.8 First i-mode protocol stacks.

cookies are not sent to the mobile terminal, and the user can switch mobile terminals and still have access to the network-stored cookies.

The WAP specific HTTP headers that are used between the mobile terminal browser and the cookie proxy are *X-Wap-Proxy-Cookie* and *X-Wap-Proxy-Set-Cookie*. The *X-Wap-Proxy-Cookie* is sent in a HTTP request from the mobile terminal browser to the cookie proxy and specifies whether the proxy should cache and manage cookies. The *X-Wap-Proxy-Set-Cookie* is sent by the proxy to the mobile terminal browser in an HTTP response, following receipt of a *Set-Cookie* from the content server, to indicate status of a session, whether cookies are sent on behalf of the terminal browser or whether an error was detected.

2.3 The i-mode architecture

The i-mode architecture developed by NTT DoCoMo in Japan was designed from the start to be closely aligned with Internet standards so as not to require content developers to learn new languages and tools for the wireless Internet [13]. This meant using a subset of HTML, namely, Compact HTML (cHTML) for Web content development, and using HTTP for the transmission of that content.

The i-mode service released in Japan in 1999 operated first on a proprietary DoCoMo transport layer (TL) on top of a Personal Digital Cellular – Packet (PDC-P) network with the protocol stacks shown in Figure 2.8 [13]. The TL layer is optimized for wireless transmissions and PDC-P operates on top of the time division multiple access (TDMA) PDC network introduced by NTT DoCoMo in 1991 as a replacement for the earlier analog networks.

With the OMA's adoption of the WAP 2.0 standards, NTT DoCoMo has committed to support these as well [14]. The more recent i-mode protocol stacks are shown in Figure 2.9. As in the WAP 2.0 architecture, a wireless-profiled version of TCP is supported. Secure communication is achieved with a Secure Socket Layer (SSL) connection

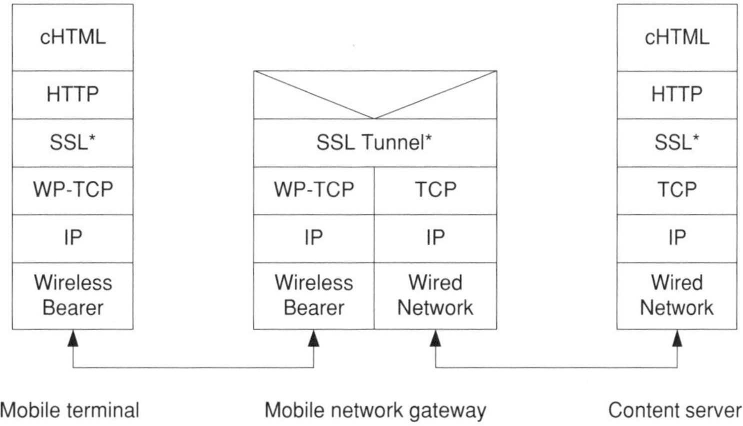

Figure 2.9 i-mode WAP 2.0-compatible protocol stacks.

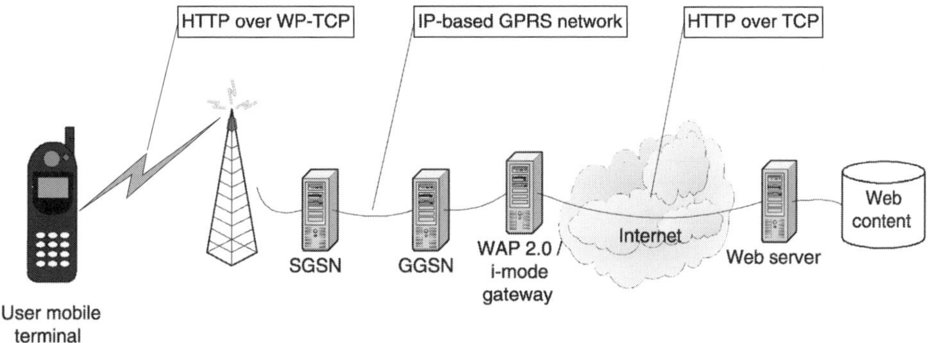

Figure 2.10 WAP 2.0 and i-mode over GPRS.

between mobile terminal and content server, and the data packets are tunneled through a mobile network gateway.

2.4 WAP 2.0 and i-mode over GPRS

As seen in the previous sections, both the WAP 2.0 and i-mode architectures have converged towards a common Internet-based architecture. Web page content is transported between mobile terminal and content server using the Internet's HTTP over TCP, with some minor enhancements for over-the-air transport.

Both the WAP 2.0 and i-mode data services have been implemented on top of the GSM cellular service offered by several European mobile operators. In this case, the wireless bearer is GSM's GPRS mobile packet data network [15]. The

WAP 2.0 / i-mode gateway server connects to a gateway GPRS support node (GGSN) which acts as the interface between the GPRS mobile network and the Internet (see Figure 2.10). The GGSN can allocate IP addresses to mobile terminals that wish to communicate with the Internet. It also maps between a terminal's IP address and the terminal's international mobile subscriber identity (IMSI). The GGSN connects to a serving GPRS support node (SGSN) which tracks the location of mobile terminals within its service area and is responsible for the delivery of data packets from and to the mobile terminals.

REFERENCES AND FURTHER READING

[1] WAP Forum, *WAP: Wireless Internet today*. (Oct. 1999).

[2] WAP Forum, *WAP Architecture*. WAP-210-WAPArch-20010712. (http://www.openmobilealliance.org/tech/affiliates/wap/wapindex.html, Jul. 12, 2001).

[3] WAP Forum, *WAP Push Architectural Overview*. WAP-250-PushArchOverview-20010703-a. (http://www.openmobilealliance.org/tech/affiliates/wap/wapindex.html, Jul. 3, 2001).

[4] WAP Forum, *Push Access Protocol*. WAP-247-PAP-20010429-a. (http://www.openmobilealliance.org/tech/affiliates/wap/wapindex.html, Apr. 29, 2001).

[5] WAP Forum, *Push OTA Protocol*. WAP-235-PushOTA-20010425-a. (http://www.openmobilealliance.org/tech/affiliates/wap/wapindex.html, Apr. 25, 2001).

[6] WAP Forum, *WAP Service Loading*. WAP-168-ServiceLoad-20010731-a. (http://www.openmobilealliance.org/tech/affiliates/wap/wapindex.html, Jul. 31, 2001).

[7] WAP Forum, *WAP Service Indication*. WAP-167-ServiceInd-20010731-a. (http://www.openmobilealliance.org/tech/affiliates/wap/wapindex.html, Jul. 31, 2001).

[8] WAP Forum, *WAP 2.0 Technical White Paper*. (Aug. 2001).

[9] WAP Forum, *Wireless Profiled TCP*. WAP-225-TCP-20010331-a. (http://www.openmobilealliance.org/tech/affiliates/wap/wapindex.html, Mar. 31, 2001).

[10] WAP Forum, *Wireless Profiled HTTP*. WAP-229-HTTP-20010329-a. (http://www.openmobilealliance.org/tech/affiliates/wap/wapindex.html, Mar. 29, 2001).

[11] D. Kristol and L. Montulli, *HTTP state management mechanism*, RFC 2109. IETF (Feb. 1997).

[12] WAP Forum, *HTTP State Management Specification*. WAP-223-HTTPSM-20001213-a. (http://www.openmobilealliance.org/tech/affiliates/wap/wapindex.html, Dec. 13, 2000).

[13] T. Natsuno, *i-mode Strategy* (Chichester, England: John Wiley and Sons Ltd, 2003).

[14] NTT DoCoMo, *i-mode Service Guideline, vl.2.0*. (Mar. 4, 2002).

[15] C. Bettstetter *et al.*, GSM phase 2+, general packet radio service (GPRS): Architecture, protocols, and air interface. *IEEE Comm. Surveys*, **2**:3 (1999).

R. G. Mukhtar *et al.*, Efficient Internet traffic delivery over wireless networks. *IEEE Comm. Mag.*, (Dec. 2003), 46–53.

3 Wireless terminals and wireless content

Current mobile terminals provide a large gamut of capabilities in an ever decreasing size and weight form factors. Today's mobile phones offer crisp color screens and versatile applications. In this chapter we review some of the most recent mobile phone capabilities, with example models from Nokia, Sony Ericsson, and NTT DoCoMo. We proceed with a description of browser-based applications, and the associated markup languages used in the creation of wireless content. The described markup languages include WML, Compact HTML, and XHTML Mobile Profile. We then conclude with an elaboration of the browser evolution with sample screens from Openwave Systems, followed by a description of the content push feature, and the capabilities offered by multi-modal browsers.

3.1 Recent mobile terminals

Most available mobile terminals offer color screens that, while still small compared to their desktop counterparts, enable the mobile user to have a far more enjoyable experience of reading text and watching pictures and videos. This is a far cry from the not so far ago days when the mobile Internet relied on screens with a few lines of black and white text and simple bitmap graphics.

In parallel with the hardware advances in the client terminals, new software capabilities were developed to ease the generation of wireless content. Initially, mobile browser markup languages that displayed wireless content were designed specifically to cater for the needs of small and resource-constrained terminals. As the processing and memory capabilities of mobile terminals improved, and the speed of data transmission over the air increased significantly, the industry moved towards a convergence of the mobile and wired Internet standards and tools. Rather than relying on tools and standards that are specialized to the wireless domain, why not leverage the same software capabilities offered in the wired domain to develop wireless systems?

Some recent model phones from Nokia, Sony Ericsson, and NTT DoCoMo, are described in the following. The new phone capabilities exhibit the dramatic transformation of mobile phones from voice communication tools to mobile terminals that

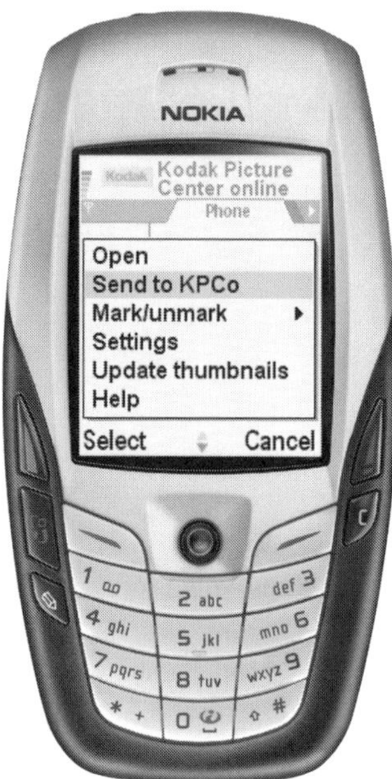

Figure 3.1 Nokia 6600 phone.
(Source: © 2004 Nokia. All rights reserved.)

provide for information display, home control, entertainment, and secure transaction handling. Some phone manufacturers have also embedded GPS receivers in some of their phone models to facilitate automatic location detection. Kyocera, for example, has included such receivers in its 2300 series phones since 2002.

Nokia series 60

The Nokia series 60 phones [1], with the 6600 model shown in Figure 3.1, provide a relatively large color screen with a size of 176 × 208 pixels. Supported connectivity includes infrared, USB, and Bluetooth. The phone has a built-in camera and an external memory card. Symbian OS is the phone's operating system, and both Symbian and Java applications are supported. Speaker dependent voice commands are enabled, allowing for hands-free operation. Users can add voice tags to applications, profiles and contacts, and activate each by voice. The included browser supports the latest release of the HTML language, HTML 4.01.

Sony Ericsson P800

The Sony Ericsson P800 phone [2], shown in Figure 3.2, includes a keypad that can be flipped open to expose a large 208 × 320 pixel touch-sensitive color screen with a stylus used for selections and text input. The large screen allows the user to write messages that

(a) Keypad closed (b) Keypad flipped open

Figure 3.2 Sony Ericsson P800 phone: (a) keypad closed; (b) keypad flipped open. (Source: © Sony Ericsson. All rights reserved.)

are interpreted by hand writing recognition software. The P800 supports the Symbian OS, can run Java applications, and includes an embedded digital camera, a video player, and an MP3 player. The internal memory is 12 MB, and an external memory card is available too. Local connectivity is enabled through Bluetooth, infrared, and USB. Finally, the browser supports WAP 2.0 and compact HTML (cHTML).

NTT DoCoMo 505i series

Figure 3.3 shows a 505i series handset offered by NTT DoCoMo [3]. The 505i features include a large QVGA color screen (240 × 320 pixels) with different font sizes so that the user can see 6, 10, or 11 lines of text. The phone is equipped with an infrared port that can be used to control home devices such as VCRs. A built-in camera allows 1.3 mega pixels resolution or reading 2D bar codes. Pictures, or other software, can be stored on a memory card that can be inserted in the phone. Finally, a fingerprint sensor on the phone enables secure transactions.

3.2 Mobile applications

Web information access is supported by a software system that is distributed across a number of network elements. A program running on the user's mobile terminal interacts with the network infrastructure Web server to retrieve the requested data. The program on the mobile terminal can be a full-fledged application with a proprietary user

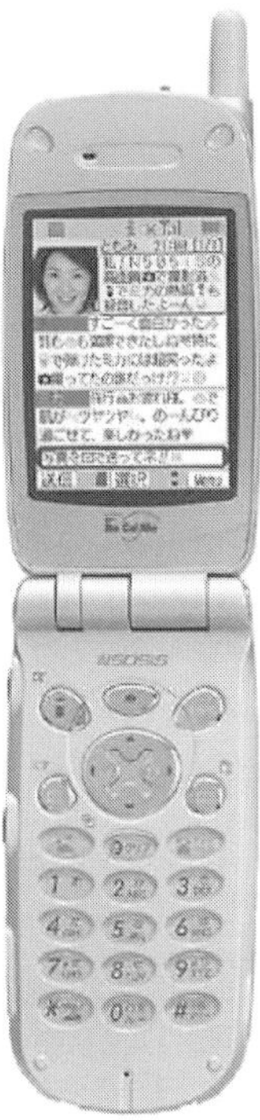

Figure 3.3 NTT DoCoMo 505i handset.
(Source: © 2003 NTT DoCoMo [3]. All rights reserved.)

interface. An example of such an application is a program that displays traffic maps. In this case, the system developer needs to program the application to run on top of the mobile terminal's operating system (OS). The OS provides the developer with an API to access services such as user interface, communication, memory management, etc.

There are a relatively large number of mobile operating systems ranging from proprietary OSs developed by mobile phone manufacturers to Qualcomm's BREW, Symbian's OS, Sun's J2ME, Microsoft's Windows CE, etc. Developers will usually have to redesign an application for a different OS when they want to port the application across different mobile terminal types, and application deployment requires that the application be downloaded to the respective client mobile terminals. Finally, a peer application is required on the server side to generate responses to the client requests.

An alternative to developing a mobile-OS dependent application for Web information access is to leverage a mobile terminal browser. Most mobile terminals support browsers that are also referred to as micro-browsers. In this case, the development effort shifts to the network server side. The developer can prepare network-stored *static Web pages* that are displayed by the terminal's browser. Better still, the content on the server can be generated dynamically by a network-based application that takes into account the request parameters and other elements of context such as the mobile terminal hardware and software characteristics, user preferences, and environment considerations.

The server would typically issue queries to content databases to retrieve the requested data. In this case, the server sends to the mobile terminal what is referred to as *dynamic content* that consists of the retrieved data augmented with appropriate markup tags. The chief advantage of this approach is that no program development is required on the mobile terminal. Consequently, any application changes are done centrally on the server side, and there is no need to perform application downloads to client mobile terminals. The next section describes the markup languages used to annotate wireless content displayed by mobile terminal browsers.

3.3 Markup languages for wireless content

The term *markup* refers to a set of tags that are affixed to data content. Markup can be used either to specify formatting directives that affect the content display or to attach semantics to the data as is done with XML. Typically, markup is used to refer to both the tags and the associated content. The hypertext markup language (HTML) is widely used on the wired Internet for formatting purposes, with the latest version being HTML 4.01 [4], now superseded with XHTML [5]. Corresponding markup languages were defined for mobile terminals, taking into account the limited screens, limited memory and processing capabilities, and the transport bandwidth constraints. These latter languages include HDML, WML, cHTML, and XHTML Mobile Profile.

3.3.1 Markup language history

Unwired Planet, which became Phone.com in 1999 and then Openwave Systems in 2000, developed in 1997 its handheld device markup language (HDML) and an associated mobile phone browser for rendering this markup. The HDML specifications were submitted that year to the W3C [6], and the browser was prevalent on most browser-based phones sold at that time. In June 1997, a few telecommunications companies that included Ericsson, Motorola, Nokia, and Unwired Planet founded the Wireless Application Protocol (WAP) Forum which designed the wireless markup language (WML) [7] as a standard successor to HDML.

On the other side of the Pacific ocean, Access [8] in Japan developed the Compact HTML (cHTML) language [9] and the Compact NetFront micro-browser. Access, jointly with other Japanese companies, submitted the cHTML specifications to the W3C organization in 1998. cHTML uses a subset of the HTML 4.0 tags.

The W3C is a major contributor to the definition of markup languages, and defined in 2000 the XHTML markup language to express HTML in XML syntax. The W3C is striving to combine the wired and wireless Web and produced in 2000 the XHTML Basic

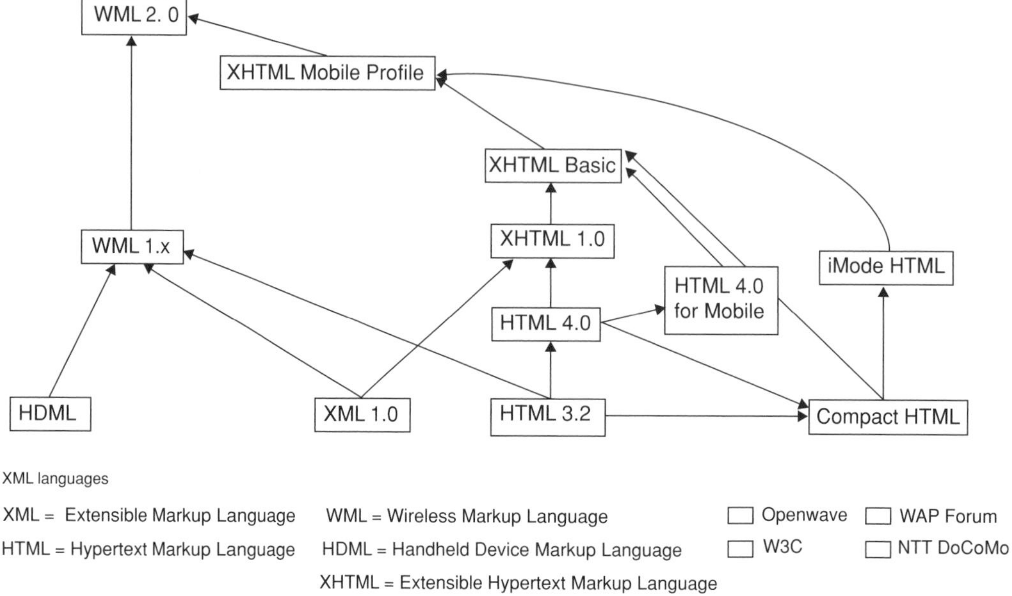

Figure 3.4 Evolution of mobile markup languages.
(Source: © 2003 Nokia [13]. All rights reserved.)

recommendation [10], a subset of XHTML that is suitable for mobile terminals. The WAP Forum extended this with additional presentation capabilities, and generated in 2001 the XHTML Mobile Profile (MP) specification [11]. Mobile phone browsers that support XHTML MP were released in the following year, and, for backward compatibility with prior content, many browsers support previous markups such as WML 1.3.

In 2001, the WAP Forum specified WML 2.0 [12]. The WML 2.0 document type is a strict superset of the XHTML Basic (including XHTML MP) document type. It proposes extensions to XHTML Basic for WML 1.x compatibility, creating a document type that is used for achieving backward compatibility with services and applications written using WML 1.x. Awaiting full approval of these extensions by the WAP Forum, the official mark-up languages for WAP 2.0 are XHTML MP and WML 1.3.

Some recent mobile phone models, for example, the Nokia 60 platform, provide support for HTML 4.01 as well as XHTML MP. The HTML support includes image maps, background images, frames, *<meta>* tags and *<object>* tags. The evolution of mobile markup languages to XHTML MP and WML 2.0 is shown in Figure 3.4. The WML, cHTML, and XHTML MP markup languages are described in more detail in the following sections.

3.3.2 WML 1.x

The wireless markup language (WML) was defined by the WAP forum with the 1.3 version released in February 2000 [7]. All information in WML is organized into a collection of *cards* and *decks*. Cards specify one or more units of user interaction (for example, a choice menu, a screen of text, or a text entry field). A user navigates through a series of WML cards to make choices, read displayed content, or enter information. Cards are grouped together into decks. A WML deck is similar to an HTML page, in that it is identified by a URL and is the unit of content transmission to a mobile terminal to prevent roundtrips to the network between cards.

WML includes support for navigation between cards and decks. The *<do>* element provides a general mechanism for the user to act upon the current card, and is mapped to a unique user interface *widget* that the user can activate. For example, the widget mapping may be to a graphically rendered button, a soft key, or a function key. For example, the WML deck in Listing 3.1 consists of two cards that enable the user to select a restaurant item and then see related information on the type of meals served.

```
<?xml version="1.0"?>
<!DOCTYPE wml PUBLIC "-//WAPFORUM//DTD WML 1.3//EN"
  "http://www.wapforum.org/DTD/wml13.dtd">
<wml>
<card>
  <do type="accept" label="Meals">
    <go href="#displaymeals"/>
  </do>
  <p>
   Choose a Restaurant:
   <select name="choice" value="meals">
     <option value="Breakfast,Brunch">Eggs Port</option>
     <option value="Lunch,Dinner">Hot Tex</option>
     <option value="Dinner,Late-night">Capricio</option>
     <option value="Breakfast,Lunch,Coffee">Pancake
       House</option>
   </select>
   </p>
 </card>

 <card id="displaymeals">
   <p>
     $choice
   </p>
 </card>
 </wml>
```

Listing 3.1 WML deck with selection list

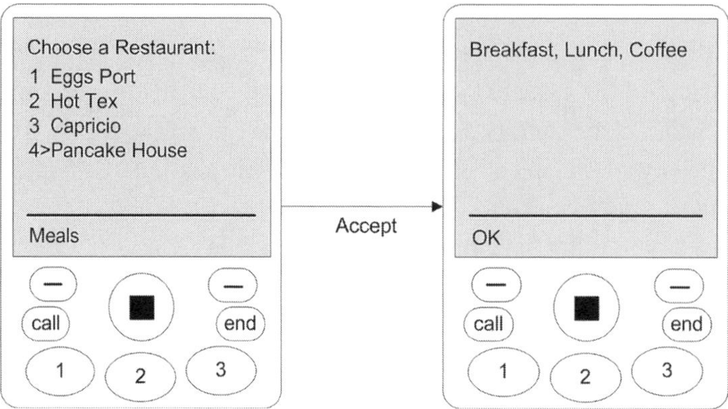

Figure 3.5 Restaurant selection screens.

WML provides for variables that can be referenced across cards and decks for a given micro-browsing session. A variable is referenced by prefixing its name with the $ sign. In Listing 3.1, upon selecting one of the displayed restaurants, the variable *choice* is set with a corresponding *value* of the list of meals served. After the user hits the *Meals* key, the card identified in the <*go*> element with the *#displaymeals* relative URL is displayed. This latter card displays the assigned value of the *choice* referenced variable, in this case the meals served by the *Pancake House* restaurant (Figure 3.5).

Navigation between decks is supported with the same <*go*> element as illustrated in Listing 3.2. In this example, the user inputs his or her name for a restaurant reservation, and this information is sent to the URL specified in the <*go*> element. A new deck will be returned to the mobile terminal to complete the restaurant reservation, for example, to enter the number of guests, desired time, and user's phone number.

```
<?xml version="1.0"?>
<!DOCTYPE wml PUBLIC "-//WAPFORUM//DTD WML 1.3//EN"
  "http://www.wapforum.org/DTD/wml13.dtd">
<wml>
<card>
  <do type="accept" label="Reserve">
    <go href="http://foo.com/servlet/reserve.servlet?
      guest=?(guest)"/>
  </do>
  <p align="center">
    <img src="restaurantlogo.wbmp" alt="Restaurant"/>
  </p>
  <p>
    Enter Reservation Name:
```

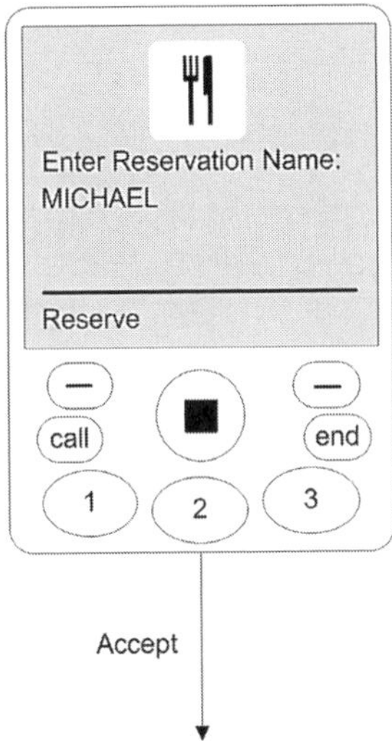

HTTP GET Request:
http://foo.com/servlet/reserve.servlet?guest=MICHAEL

Figure 3.6 Restaurant reservation screen.

```
    <input name="guest" format="10A"/>
  </p>
</card>
</wml>
```

Listing 3.2 WML input card

As one of the major design considerations of WML is efficient transmission over low-bandwidth networks, the WML specification defined Wireless Bit Map (WBMP), an efficient graphic format for black-and-white images. This does not preclude the use of other graphic formats such as GIF and JPEG provided they are supported by the browser. Corresponding images are defined with the *<img>* tag, shown in Listing 3.2, and two mandatory attributes, the *src* attribute to indicate the image URL, and the *alt* attribute that specifies output text if the image cannot be displayed.

After the user hits the *Reserve* key in the screen in Figure 3.6, an HTTP GET request with the guest's name is sent to the URL specified in the *<go>* element.

The WAP Forum has produced a companion standard to WML that specifies *WMLScript*, a procedural scripting language designed to enhance WML, similar to the way JavaScript enhances HTML pages [14]. With WMLScript, the programmer can define functions that manipulate basic data types such as Booleans and integers, and include flow control statements, assignments, logical, arithmetic, and comparison operators.

50	58991	E66F		Restaurant	Black
51	58992	E670		Cafe	Green
52	58993	E671		Bar	Purple
53	58994	E672		Beer	Orange
54	58995	E673		Fast food	Orange

Figure 3.7 Restaurant services emoji icons.
(Source:)

In the WAP 1.x specifications, WML text and WMLScript programs are binary encoded to produce compressed versions of the full text format. This requires a WAP gateway server that translates between the text format and the binary code for over-the-air transmission. The client browser also needs to support bytecode interpreters for the binary encoded WML and WMLScript.

3.3.3 Compact HTML

Compact HTML (cHTML) is the markup language used in NTT DoCoMo's i-mode wireless Internet system, and is also referred to as iHTML [15], [9]. cHTML is a subset of HTML that does not include the more complex formatting functions such as tables, frames, image maps, and cascading style sheets. Although GIF images are supported, JPEG is not supported, and neither is script programming. i-mode browsers that accept cHTML content use a two-byte character encoding scheme called Shift-JIS [16] in Japan while Unicode encoding is used outside Japan [17].

Since cHTML is a subset of HTML, it does not support the concepts of card and deck of WML. Each hyperlink invocation downloads the document referenced by the link as one page that could be scrolled on the mobile terminal if it is longer than the screen's size. The cHTML language supports extensions to facilitate the generation of screen displays for mobile terminals. These extensions include an *accesskey* attribute (also defined in HTML 4) that is contained in the *anchor* HTML element, and indicates the key that can be pressed to activate the associated link. Another extension is the use of the *emoji* icon symbols which display as unique character-sized graphics (12×12 pixels) that save screen space. These icons convey varied notions such as weather, transportation means, restaurant facilities, sports, human feelings, and plain numbering. As of early 2004, NTT DoCoMo had defined 236 such icons. For example, restaurant services icons are shown in Figure 3.7.

Each emoji icon is referenced in the cHTML code by preceding its decimal code with the symbols *&#*. For example, the Café icon (in slot 51) is referenced by **. The cHTML code in Listing 3.3 generates a screen with a list of restaurant services in the mobile user's vicinity. This code shows how the *accesskey* attribute is used to associate list options with keypad keys, and how to include emoji graphic codes.

```
<! DOCTYPE html PUBLIC "-//W3C//DTD Compact HTML 1.0
   Draft//EN">
<html><head><title>Restaurants</title></head>
<meta <http-equiv="Content-Type" content="text/html;
  charset=SHIFT_JIS">
<body>
  <div align="center">Restaurants</div>
  <div align="center">In Your Area</div>
  <hr/>
  <div align="center">Press Key to Select</div>
  &#59106; <a href="http://www.eggsport.com/index.htm"
  accesskey="1">
  Eggs Port</a> &#58991;<br/>
  &#59107; <a href="http://www.hottex.com/index.htm"
  accesskey="2">
  Hot Tex</a> &#58995;<br/>
  &#59108; <a href="http://www.capricio.com/index.htm"
  accesskey="3">
  Capricio</a> &#58991;<br/>
  &#59109; <a href="http://www.pancakehouse.com/index.htm"
  accesskey="4">
  Pancake House</a> &#58992;<br/>
  <hr/>
</body>
</html>
```

Listing 3.3 i-mode cHTML of restaurant list

As the i-mode screen in Figure 3.8 shows, emoji icons that are descriptive of the respective restaurant services are displayed next to each item on the list.

3.3.4 XHTML Mobile Profile

XHTML 1.0 [5] is a reformulation of HTML 4 in XML that was produced by the W3C in 2000. As such it includes powerful features for authoring Web content, but its design does not accommodate small devices, for example cell phones that have limited

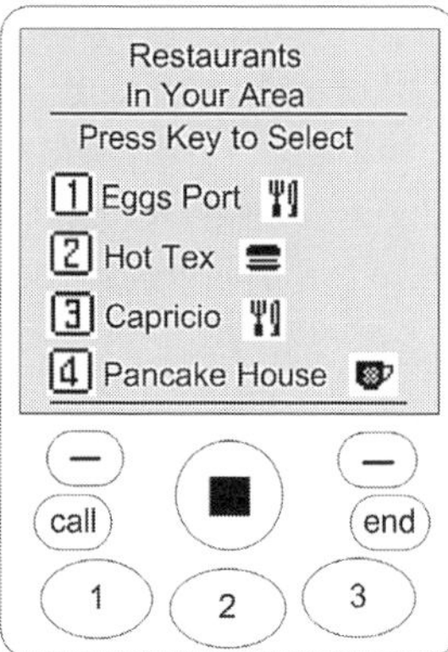

Figure 3.8 i-mode screen of restaurant list.

computing and display capabilities. Hence, the motivation for XHTML Basic [10] to provide an XHTML document type that can be used for mobile terminals and that is rich enough to be used for simple content authoring.

While XHTML Basic supports forms and basic tables, not including embedded tables, more complex features such as scripting, frames, and the HTML *<style>* element, are not supported. The rationale for not supporting these capabilities is that the CPU, memory limitations, and small screens of mobile terminals are not conducive to support resource-demanding features.

The W3C defined a Cascading Style Sheet (CSS) language of formatting directives to get preferred font, color, and layout effects associated with content elements. External CSS can be used jointly with XHTML documents to create desired presentations. The WAP Forum defined a corresponding WAP CSS (WCSS) specification with mobile terminal extensions [18], and the W3C also issued a candidate recommendation of CSS tailored to the needs and constraints of mobile terminals [19]. The WAP Forum's extensions include the *marquee* element property that enables horizontal text scrolling, the *input* property to specify the format of user input in forms, and the *accesskey* property to enable activations of elements with mobile terminal keys.

The external WCSS are processed in order of appearance, hence the name cascading, and are invoked with *<link>* elements that are included in a document's *<head>* element. An example sequence of WCSS invocations is shown in Listing 3.4.

```
[1]  <link rel="stylesheet" href="corporate.css" type="text/css">
[2]  <link rel="stylesheet" href="techreport.css" type="text/css">
```

Listing 3.4 Cascading style sheets sequential invocation

The XHTML Mobile Profile (XHTML MP) specification [11] produced by the WAP Forum is a strict subset of XHTML that extends XHTML Basic with some enhanced

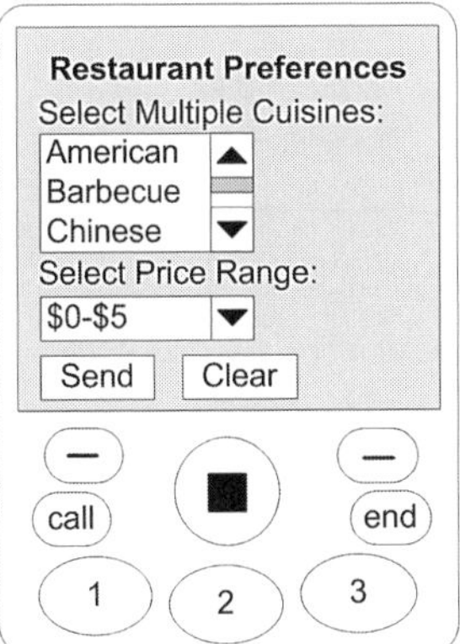

Figure 3.9 Restaurant preferences input form.

functionality, which includes additional presentation elements, internal style sheets, and inline styling. An internal style sheet is a collection of style directives located within the markup document and enclosed by *<style>* element tags. Inline styling can be specified for single elements using the *style* attribute to describe a particular format, for example, centered text. The markup in Listing 3.5 shows an internal style sheet for centering all the document's paragraph elements, and an inline style that left aligns a specific paragraph in the same document, since inline styling takes precedence over other styling directives.

```
<head>
  <style type="text/css">
     p {text-align: center;}
  </style>
<head>
...
<p style="text-align: left">...</p>
```

Listing 3.5 Internal style sheet and inline styling

The XHTML Mobile Profile markup in Listing 3.6 is used in a profile builder application. The markup generates an input form in which the mobile user can specify his restaurant preferences. The first *<select>* element allows the mobile user to specify multiple favored cuisines, and with the second *<select>* element the user can specify a preferred price range. The *<input>* elements at the end define screen buttons for sending and clearing the form.

Included in the markup is an internal style sheet, specified with the *<style>* element, that assigns the blue color to all option elements in the menus, and adds padding to the input buttons to increase their size. The displayed input form is shown in Figure 3.9.

```
<?xml version="1.0"?>
<!DOCTYPE html PUBLIC "-//WAPFORUM//DTD XHTML Mobile
1.0//EN" "http://www.wapforum.org/DTD/xhtml-mobile10.dtd">
<html>
  <head>
    <title>Restaurant Preferences</title>
    <style type="text/css">
      option {color:blue}
      input {padding-left:5}
      input {padding-right:5}
    </style>
  </head>
  <body>
    <h4>Restaurant Preferences</h4>
    <form action="RestaurantsCo.com" method="get">
      Select Multiple Cuisines:<br/>
      <select name="Choice1" multiple="multiple">
        <option>American</option>
        <option>Chinese</option>
        <option>French</option>
        <option>Italian</option>
        <option>Mexican</option>
      </select><br/><br/>
      Select Price Range:<br/>
      <select name="Choice2">
        <option>$0–$5</option>
        <option>$5–$15</option>
        <option>$15–$30</option>
        <option>$30 and up</option>
      </select><br/><br/>
      <input type="submit" name="Submit" value="Send"/>
      <input type="reset" name="Reset" value="Clear"/>
    </form>
  </body>
</html>
```

Listing 3.6 XHTML Mobile Profile of input form

One of the major benefits of using XHTML MP is that, except for minor adaptations, the same tools used for wired Internet content can be used for wireless content development. The other advantage is that external style sheets, in the form of WCCS, specify separately from the content all formatting aspects, including positioning, fonts, text

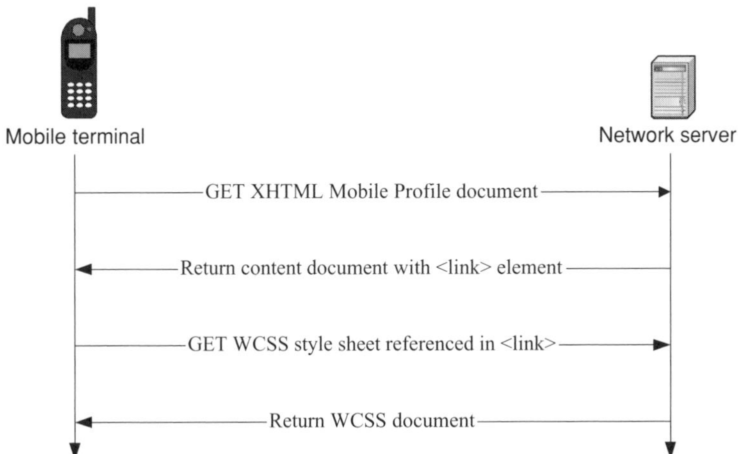

Figure 3.10 Separate content and WCSS documents download.

attributes, borders, and margin alignment. A separate WCSS can be defined for each terminal type to enhance its content presentation. This is unlike WML 1.x markup where formatting instructions are intermixed with the content. The order of content and WCSS document downloads is shown in Figure 3.10.

3.3.5 WML 2.0

The WAP Forum defined WML 2.0 [12] with the goal of creating a document type that extends the syntax and semantics of XHTML Mobile Profile and CSS Mobile Profile with the unique semantics of WML 1.x for backward compatibility purposes. Client mobile terminals that host browsers that are WML 2.0 compliant will not require dual browsers supporting both XHTML Mobile Profile and WML 1.x to leverage WML 1.x features. So far, WML 2.0 has not been approved by the WAP Forum, and the official languages of WAP 2.0 are WML 1.3 and XHTML MP.

Example concepts missing in XHTML that were defined in WML 1.x are cards, for representing individual interactions with the user, and variables, for passing values between cards. These concepts were therefore reintroduced in WML 2.0, and the corresponding elements are *<card>* for specifying a document fragment, and *<setvar>* and *<getvar>* for managing variables.

3.4 Browser evolution

3.4.1 Browser rendering

As the markup languages have evolved their support for better graphics, the mobile terminal browsers have advanced correspondingly. The browser screens of the Yahoo!

Figure 3.11 Openwave browser evolution.
(Source: © 2003 Openwave Systems. All rights reserved.)

home page in Figure 3.11 show the progression of browser rendering capabilities for the Openwave Systems browser products. Browser version 4.1 supports WML 1.1 and is a text-only renderer with limited graphics support such as WBMP. Version 5 supports WML 1.3 with extensions defined by GSM M-services [20] that provide HTML-like form capabilities to WML 1.3. Version 5 includes color support and supports the JPEG, PNG, GIF, and animated GIF graphic formats, in addition to BMP and WBMP. The unsolicited delivery of content referred to as *content push* is also supported in version 5. Finally, version 6.2 supports both a subset of HTML as well as XHTML Mobile Profile with WCSS and delivers innovative new features such as progressive layout, allowing users to interact with web pages before they are fully loaded, and new table algorithms to improve the rendering of HTML content.

Other companies, for example, Opera and Access, provide support for HTML 4.01 in their micro-browser offerings. The Opera micro-browser 7 and the Access micro-browser NetFront v.3.1 were designed to enable desktop-style browsing on small screen terminals. These browsers use renderers that reformat content for the desktop to display on small screens without any horizontal scrolling. This is done by restructuring a page content to fit in a vertical column with a width equal to the width of the mobile terminal screen.

3.4.2 Content push

As described in the previous chapter, the WAP architecture provides support for unsolicited content delivery to mobile terminals, referred to as *content push* [21]. This action is initiated by a network server, and there are two possible content push services, a *service indication* and a *service load*. In both services, a URI that points to the pushed content is sent to the browser. The browser initiates a corresponding HTTP GET, or

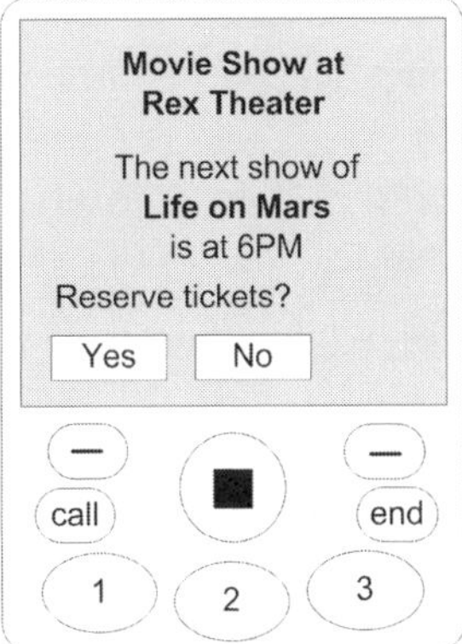

Figure 3.12 Push notification of movie show.

WSP GET for WAP 1.x, to retrieve the content. When a mobile terminal receives a *service indication* push message, a short text description is typically displayed on the terminal's screen. The user is given the choice of immediate or delayed display. If the received push message is a *service load* message, then the user's settings will determine if the page is automatically loaded and displayed, or if it is loaded to the mobile terminal's cache for later display.

Support for push services is a standard feature in many released mobile terminals. The novel aspect in the push service is the new interaction paradigm that is offered to mobile users. Rather than sending the mobile user a notification message in the form of an SMS message or an email, the user is presented with a new screen that includes control buttons that the user can activate as desired. In the former approach, the user would have to type back a response message, whereas in the content push approach all the user is typically requested to do is select an option or click a button. For example, a tourist user can be presented with a pushed screen that indicates that a movie he or she wants to see will be playing shortly in a nearby theater. The user is presented with the option to reserve tickets as shown in the screen of Figure 3.12.

3.4.3 Multi-modal browsers

Multimodal technology allows the interchangeable use of multiple forms of input and output, such as voice commands, keypads, stylus, gestures, and vision in the same interaction. Applications that leverage multimodal interactions can benefit users that require hands and eyes free operation, for example, drivers that request navigation instructions or workers on the factory floor that require inventory information by voice. There may even be situations where interacting with one device, for example a cell phone, causes output to appear on another one, for example a PDA. The benefit of multimodal interfaces is their ability to switch between different input modes depending on the current context and physical environment of the mobile user. This adaptive capability makes them particularly appealing as context dependencies play a major role in the design of wireless Web information services.

The W3C multimodal interaction working group (WG) [22] was chartered to work on developing standards for synchronization across modes and devices. Different approaches are considered by this WG to develop the required standards. The WG strives to reuse existing W3C standards including VoiceXML [23] and Synchronized Multimedia Integration Language (SMIL) [24]. VoiceXML is an XML-based dialog design language to build conversational applications, and SMIL is a markup language for synchronizing the display of various types of media. For example, SMIL markup can ascertain that audio clips are played at the same time as corresponding images are displayed on the terminal's screen. The evaluated approaches for multimodal technology include the ability to speech enable HTML or XHTML markup, combining XHTML and VoiceXML markup, or combining SMIL with markup for control of speech engines to support multimodal interaction.

REFERENCES AND FURTHER READING

[1] Nokia, *Series 60 Platform Product Overview.* (http://www.forum.nokia.com/main/0,6566, 010_ 40,00.html, Jan. 2004).

[2] Sony Ericsson, *P800.* (http://www.sonyericsson.com/spg.jsp?cc=us&lc=en&ver=4000& template= pp1_1_1&zone=pp&lm=pp1&pid=9940,2004).

[3] NTT DoCoMo, *505i handsets.* (http://www.nttdocomo.com/presscenter/pressreleases/press/ pressrelease.html?param[no]=218,2003).

[4] D. Raggett *et al., HTML 4.01 Specification.* W3C Recommendation (Dec. 24, 1999), http:// www. w3.org/TR/html401/.

[5] S. Pemberton *et al., XHTML 1.0: The Extensible Hypertext Markup Language – A Reformulation of HTML 4 in XML 1.0.* W3C Recommendation (Jan. 26, 2000), http://www.w3.org/TR/xhtml.

[6] T. Hyland. *Proposal for a Handheld Device Markup Language.* W3C Note (May 9, 1997), http:// www.w3.org/TR/NOTE-Submission-HDML.html.

[7] WAP Forum, *WAP WML Specificaion v1.3.* WAP-191-WML. (http://www.openmobilealliance. org/tech/affiliates/wap/wapindex.html, Feb. 19, 2000).

[8] Access Co., http://www.access.co.jp/english/company/index.html.

[9] T. Kamada. *Compact HTML for Small Information Appliances.* W3C Note (Feb. 9, 1998), http://www.w3.org/TR/1998/NOTE-compactHTML-19980209.

[10] M. Baker *et al., XHTML Basic.* W3C Recommendation (Dec. 19, 2000), http://www.w3.org/TR/ 2000/REC-xhtml-basic-20001219.

[11] WAP Forum, *XHTML Mobile Profile.* WAP-277-XHTMLMP-20011029-a. (http:// www.openmobilealliance.org/tech/affiliates/wap/wapindex.html, Oct. 29, 2001).

[12] WAP Forum, *Wireless Markup Language v2.0.* WAP-238-WML-20010911-a. (http:// www. openmobilealliance.org/tech/affiliates/wap/wapindex.html, Sep. 11, 2001).

[13] Nokia, WML to XHTML migration v2.1. *Forum Nokia* (http://www.forum.nokia.com/ nds- CookieBuilder?fileParamID=2995, Apr. 30, 2003).

[14] WAP Forum, *WMLScript Language Specification.* WAP-193-WMLScript. (http:// www.open- mobilealliance.org/tech/affiliates/wap/wapindex.html, Mar. 24, 2000).

[15] NTT DoCoMo, *How to Create An i-Mode Site v1.1.* (http://www.imode.nl/imode/gfx/attachments/How%20to%20create%20an%20i-mode%20site.pdf, Apr. 15, 2002).

[16] School of Modern Languages, Shift-JIS table. *Georgia Institute of Technology* (http://www. modlangs.gatech.edu/Programs/Japanese/s-jis.html).

[17] The Unicode Consortium, *The Unicode Standard v4.0.* (Addison-Wesley, 2003).

[18] WAP Forum, *WAP CSS Specification.* WAP-239-WCSS-20011026-a. (http://www.openmobilealliance.org/tech/affiliates/wap/wapindex.html, Oct. 26, 2001).

[19] T. Wugofski *et al., CSS Mobile Profile 1.0* W3C Candidate Recommendation (Jul. 25, 2002), http://www.w3.org/TR/2002/CR-css-mobile-20020725.

[20] GSM, *M-Services Guidelines v3.0.0.* (http://www.gsmworld.com/documents/m-services/aa35.pdf, May 31, 2001).

[21] WAP Forum, *WAP Push Architectural Overview.* WAP-250-PushArchOverview-20010703-a. (http://www.openmobilealliance.org/tech/affiliates/wap/wapindex.html, Jul. 3, 2001).

[22] W3C, *Multimodal Interaction Working Group Charter.* (2004), (http://www.w3.org/2002/01/multimodal-charter.html).

[23] S. McGlashan *et al., Voice Extensible Markup Language (VoiceXML) v2.0.* W3C Recommendation (Mar. 16, 2004), http://www.w3.org/TR/voicexml20/.

[24] J. Ayars *et al., Synchronized Multimedia Integration Language (SMIL 2.0).* W3C Recommendation (Aug. 7, 2001), http://www.w3.org/TR/smil20.

Motorola, *iDEN Devices*, http://idenphones.motorola.com/iden/developer/java_specs.jsp.

Phone.com, *UP.SDK Developer's Guide v4.0.* Phone.com (Jan. 2000).

4 User mobility and location management

In this chapter we address the topic of user location, a key element of context for mobile applications. Initially, we review the concept of IP addressing and the mobile IP standard that enables users to maintain an IP connection when roaming between networks. We then describe handset-based and network-based approaches to determine mobile user location. We follow with a description of the Open Mobile Alliance's (OMA) effort at defining standard protocols for accessing location information.

Location information is typically hierarchical, and we review hierarchical schemes for representing location information. Tree-based schemes, including R-trees and Quadtrees, are described, and we outline location tracking approaches. Moving objects databases constitute a new research area. We elaborate on how these databases address the unique modeling requirements of moving objects and support queries that can simultaneously handle space and time constraints. Finally, we describe the US E911 emergency services government mandates and the implemented solutions for detecting the location of mobile users that make emergency calls.

4.1 IP and mobility

A network element's interfaces are identified by the unique Internet Protocol (IP) addresses assigned to each interface. Typically, a mobile terminal will have just one assigned IP address, although other network elements, for example network routers, may have multiple addresses. An IP address consists of a network prefix and a host portion. For example, the IP version 4 (IPv4) address "128.32.17.25" is 32-bits long and consists of a network prefix "128.32.17" and a host portion "25". The network prefix determines the link where the host is attached, and the host portion specifies the unique host identifier on this link.

IP version 6 (IPv6) is a new version of the Internet Protocol standardized by the IETF [1], designed to eventually replace IPv4. IPv6 increases the IP address size from 32 bits to 128 bits, to support more levels of addressing hierarchy, a much greater number of addressable nodes, and simpler auto-configuration of addresses. Among the new headers defined in IPv6 is a flow label to label sequences of packets for which special handling can be performed by the IPv6 routers, such as specific management of real-time streams, for example, video.

The home network operator can manually assign IP addresses to its subscribers' mobile terminals. Alternatively, a Dynamic Host Configuration Protocol (DHCP) server [2] can dynamically assign an IP address to a terminal that attempts to connect to the network. Applications running on a network element that wish to communicate across the network will have, in addition to an IP address, a port number that uniquely identifies their connection point. For example, port 80 is the default port number to which a Web server listens for HTTP requests.

Mobile operators are strong advocates of the transition to IPv6 as the larger address space will enable each mobile terminal to have a globally unique address, particularly critical for peer-to-peer networking where this is required by each communicating end. To circumvent the lack of sufficient IP addresses, a mobile operator assigns each mobile terminal within its domain with a private IP address known only within the mobile operator's network. When the terminal communicates with the outside Internet, a network address translator (NAT) assigns the terminal a temporary public IPv4 address and keeps track of the mapping between private IP addresses and assigned public IPv4 addresses. With IPv6 and its large address space there will be no need for NAT elements. Nokia, for example, sees the transition taking place through the support of dual IPv4 and IPv6 stacks in both mobile terminals and network infrastructure elements until the time when IPv6 is dominant [3]. The other mechanism that will play a major role in the transition is tunneling, where IPv6 packets are encapsulated by IPv4 packets and are "decapsulated" at the other end of the tunnel. This tunneling enables IPv6 mobile terminals to access IPv6 services through an intermediary IPv4 Internet and relies on dual stack support at both ends of the tunnel.

4.1.1 Mobile IP packet routing

A mobile terminal that changes its connection point to the Internet may have to terminate all existing communications before reattaching, usually with a different IP address, at another point. This type of disconnection and reconnection is referred to as "nomadicity". However, when wireless mobile terminals roam between different locations they may wish to keep up their current session without resorting to restarting it whenever they change locales. This capability is referred to as "seamless mobility". Different layers of the terminal's communication protocol stack need to provide their respective support for this type of mobility. The link layer maintains an ongoing connection when moving between connection points through what is referred to in cellular networks as "handover". The IETF's Mobile IP standard ([4], [5]) defines support procedures at the next layer, the network layer, of the protocol stack.

Mobile IP enables mobility between nodes of the same or different network types; for example, between a wireless wide area network (WWAN) such as GPRS and

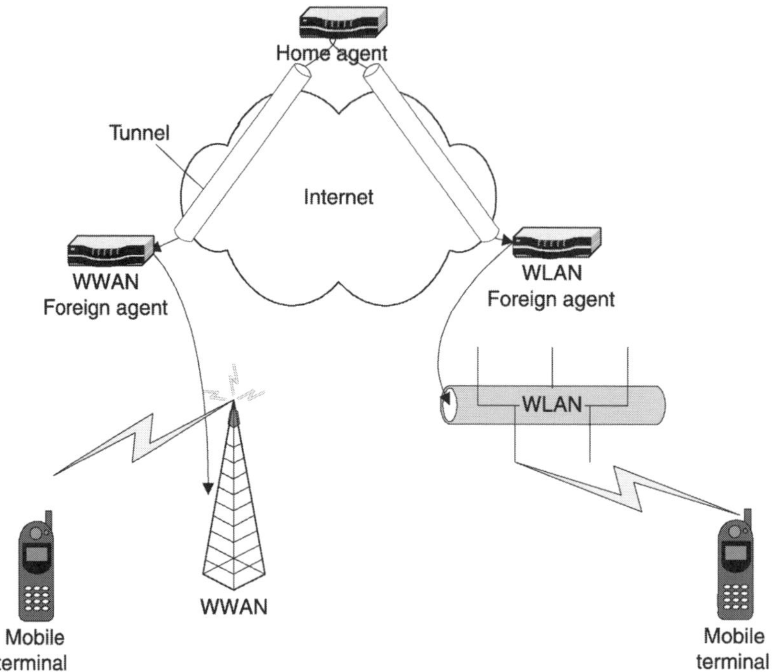

Figure 4.1 Mobile IP packet routing through foreign agents.

a wireless local area network (WLAN) such as 802.11 (see Figure 4.1). It allows mobile terminals to maintain all ongoing communications while changing their Internet connection link, without changing the mobile terminal's IP address. The Mobile IP architecture enables this mobility with two network elements: a home agent router and a foreign agent router shown in Figure 4.1. The foreign agent informs a home agent of a mobile terminal's current "care-of address", and forwards received terminal-destined packets to the connected terminal. The home agent tracks the mobile terminal's location (its care-of address), and forwards terminal-destined packets to the current terminal's care-of address at the foreign agent. A home agent is said to "tunnel" a terminal-destined packet to the foreign agent since it encapsulates the packet in an outer packet with the foreign agent's address. The foreign agent "de-tunnels" the received packet, that is, removes the envelope packet and delivers the original packet to the terminal.

4.1.2 Agent discovery and registration

Agent discovery is the process by which a roaming mobile terminal will listen to advertisement messages issued by home and foreign agents to determine if it is connected to a home or foreign link. These mobile IP messages are referred to as "agent advertisements", and the mobile terminal will typically compare the network prefix of the agent's IP address to its own home address network prefix. If they differ, then the mobile terminal will conclude that it is on a foreign link, and, in IPv4, it will acquire a care-of address directly from those addresses advertised in the agent advertisements.

If the mobile terminal does not hear any advertisement messages within a specified period, it can issue mobile IP "agent solicitation" messages. If no response is sent back, then the mobile terminal can resort to obtain an address using an assignment procedure from a dedicated DHCP server responsible for assigning Internet addresses to connecting terminals. Once a mobile terminal has acquired such an address it can use this address as a "collocated care-of address", and register it with its home agent using a mobile IP registration protocol in a similar way that a care-of address is registered by a foreign agent.

Unlike the foreign agent assignment of addresses in IPv4, in IPv6 the generation of an IP address is performed locally at the mobile terminal through a stateless address auto-configuration mechanism [6]. The stateless mechanism allows a terminal to generate its own address by combining its own interface identifier with a link's subnet identifier as specified in router advertisements on the associated link. This address is a collocated care-of address as it identifies the mobile terminal itself; as previously mentioned, such an address is also used in IPv4 when no foreign agent is available. There is, therefore, no need for a foreign agent concept in IPv6. To insure that all configured addresses are likely to be unique on a given link, a terminal can run a "duplicate address detection" algorithm as defined in [6] before assigning an address to an interface.

As mentioned above, seamless mobility is enabled with Mobile IP. For example, when a mobile user has an ongoing session with a Web content server and roams between networks, the user's terminal could acquire a new IP care-of address without interrupting the session. While an IP care-of address provides an indication of the approximate whereabouts of a mobile user, this can be at times a relatively large area. When connected to a WLAN this area could be a few hundred meters wide, and when connected to a WWAN this could be a much larger area that covers many cells. In the next section we describe approaches which provide more accurate position determination for location-based services.

4.2 Location determination

The precise location of a mobile user or of a point-of-interest (POI) on the surface of the earth is provided by a "geocode" that includes the latitude and longitude, and possibly the altitude. Geocodes can be expressed in decimal form or in nautical form

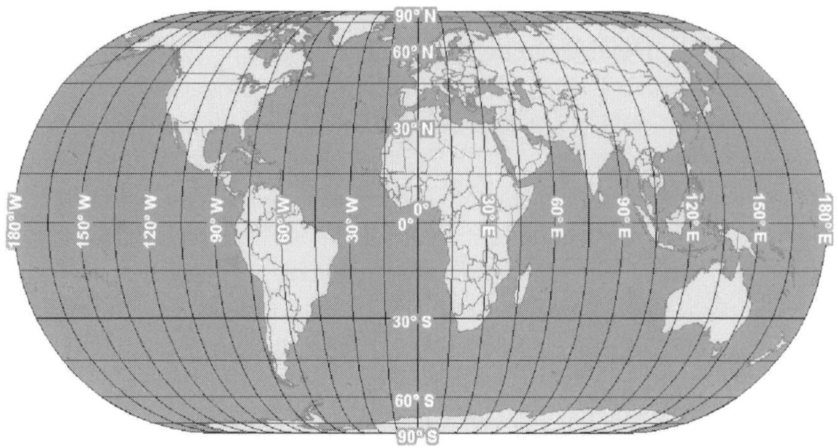

Figure 4.2 World latitude and longitude lines.
(Source: Peter H. Dana, University of Texas, Austin.)

and are shown in Figure 4.2. The Equator is at 0 degrees latitude, the North Pole at +90 degrees, and the South Pole at −90 degrees. Longitude is measured relative to the Prime Meridian at 0 degrees longitude, passing through the Royal Greenwich Observatory in London, England [7], [8]. Longitude goes East up to 180 degrees, and West to −180 degrees. The overlap of the 180 degrees East and −180 degrees West is the International Date Line passing through the Pacific Ocean.

Nautical geocodes include degrees followed by 0 to 59 minutes, followed by 0 to 59 seconds. For example, the latitude of the center of the town of Evanston, Illinois, on the shore of Lake Michigan, as provided by the US Census Bureau [9], is 42°2′47″ N by 87°41′40″ W in nautical form, and in decimal form the latitude is 42.04635 degrees and the longitude is −87.69454 degrees (see Figure 4.3).

To compute the location of a mobile terminal, a positioning system tracks and measures radio signals emitted by the mobile terminal, the neighbor cell sites, or orbiting satellites. The approach used to determine a mobile user's position is either handset-based or network-based, depending on where the radio measurements and computations are performed. Some hybrid approaches are used as well, where measurements and computations are split between the terminal and the network. The radio measurements are of signal strength, signal angle of arrival, and signal time of arrival. The first two, signal strength and angle of arrival measurements, are error prone and cannot provide accurate indications, so that the prevailing methods adopted by standards organizations and implemented by mobile operators rely on measuring signal time of arrival [10]. The 3GPP cellular standards organization has settled on three wireless location technologies for 3G networks [11], [12]. These include the cell-ID-based method, Observed Time Difference of Arrival (OTDOA), and Assisted Global Positioning System (A-GPS), and are described in the following sections.

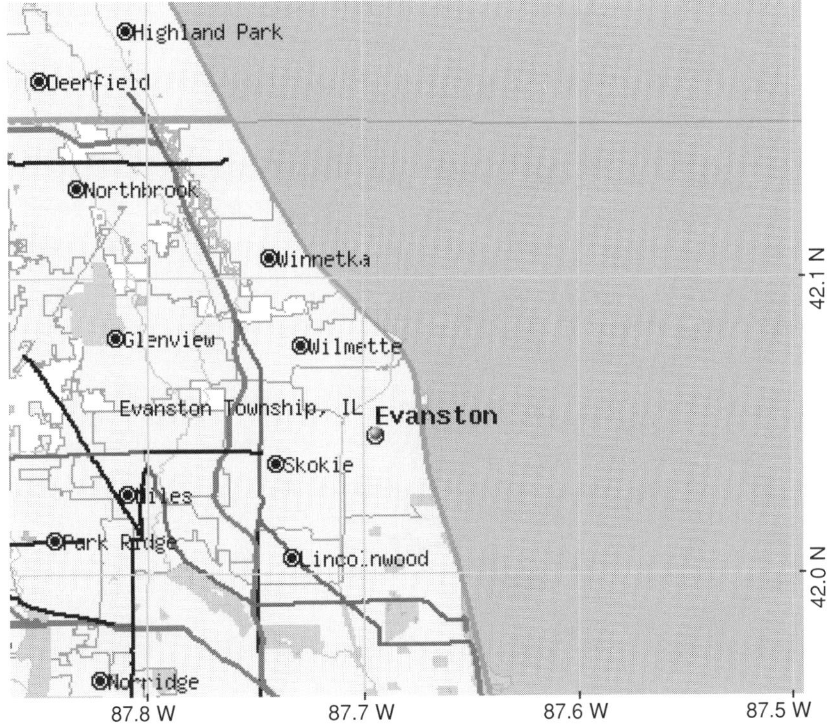

Figure 4.3 Map of Evanston, Illinois, with latitude and longitude coordinates. (Source: US Census Bureau.)

4.2.1 Handset-based position determination

Handset-based approaches for computing the position of a mobile terminal require add-on software on the terminal, and at times also special sensors, for example, satellite signal receivers.

OTDOA positioning

The handset-based Observed Time Difference of Arrival (OTDOA) method ([10], [13]) typically uses two pairs of cell sites, where some sites could be the same across pairs (see Figure 4.4). The mobile terminal measures the time difference between signal arrivals for each pair of cells. Each such measurement defines a hyperbolic locus on which the mobile terminal is at a constant time difference of arrival from the corresponding pair of cells. The intersection of the two hyperbolic loci determines the mobile terminal's position. This approach requires the mobile terminal to have good reception from at least three separate cells, which may not always be the case.

A corresponding network-based position determination approach, described in the next section, is to have the cell sites listen to the signals emitted by a mobile terminal,

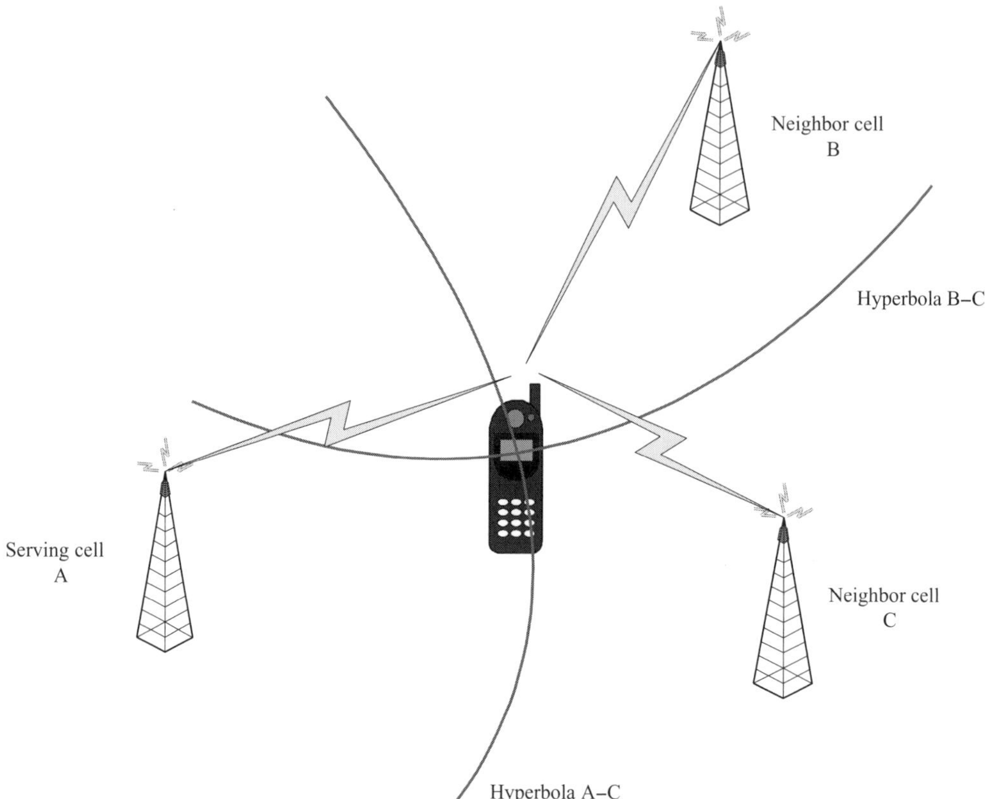

Figure 4.4 Handset-based OTDOA position determination.

record the times of arrival, and send these times to a central site for position computation. In both cases, there is a need for timing synchronization. Either the network's cell site clocks are all synchronized, or else there is a separate monitoring site that tracks the timing offset of each cell site. In the latter case, the timing offsets are conveyed to the place where the position computation is done: either the mobile terminal or a network site.

GPS positioning

The Global Positioning System (GPS) relies on a satellite navigation system funded by and controlled by the US Department of Defense ([14], [15], [16]). The first satellite was launched in 1978 and deployment was completed in 1994. There are a total of 24 GPS satellites that orbit the earth in six orbital planes, and provide coded signals that can be processed in a GPS receiver, enabling the mobile terminal receiver to compute its position to within 20 to 100 meters in unobstructed environments. Between five and eight GPS satellites are visible from any point on the Earth. A receiver requires only four GPS satellite signals to compute its position in three dimensions using a technique commonly referred to as triangulation. Signals from additional satellites, if available, are used to increase the confidence in the computation.

The GPS system started as a military system, and applied a Selective Ability (SA) mechanism that used intentional signal degradation by a time varying bias so that non-military users achieved only about 100 meters accuracy in position determination. On May 1, 2000, SA signal scrambling was turned off by the US government, enabling commercial applications to benefit from the same accuracy provided to the military. The European Space Agency is building a system similar to GPS, called Galileo [17], that will be interoperable with GPS. Galileo will include 27 operational satellites and will be completed in 2008.

GPS position is computed at the receiver by determining the distance to each of the satellites from which signals are received. Distance calculation is done by measuring the difference between the time a signal is sent and the time it is received for each of the incoming satellite signals. As errors may occur in the time difference measurements, due mainly to signal propagation delays and multipath fading (where signals bounce off surrounding objects), more accurate position computations were introduced that enable determining location to within 20 meters [15]. These methods include Differential GPS (D-GPS) and Assisted GPS (A-GPS), and rely on a fixed receiver, for example, a cell site that knows its position and can therefore estimate the bias error of each satellite. The fixed receiver forwards the bias errors to the mobile terminals so that they can be taken into account when determining position (see Figure 4.5). In addition, the cellular network's location tracking servers inform the fixed receiver about the approximate mobile terminal's location, for example, the mobile terminal's serving cell site location. The fixed receiver can then send satellite information to the mobile terminal that indicates to which satellite signals the terminal should listen to. This greatly reduces the terminal's time to determine position on start-up from a few minutes down to a few seconds [18].

The location accuracy of mobile terminals that have integrated GPS receivers is variable and could be within 50 meters as in NTT DoCoMo's F661i terminal [19]. In the Motorola i730 GPS-enabled terminal described in [20], the position determination is activated each time a user issues an emergency call. Position computation may then take 30 seconds or more, depending on the number of accessible satellites and whether the cellular network provides A-GPS.

4.2.2 Network-based position determination

Unlike handset-based approaches for computing position, network-based approaches typically do not require upgrades in the mobile terminals; however, they may require special equipment at the cell sites.

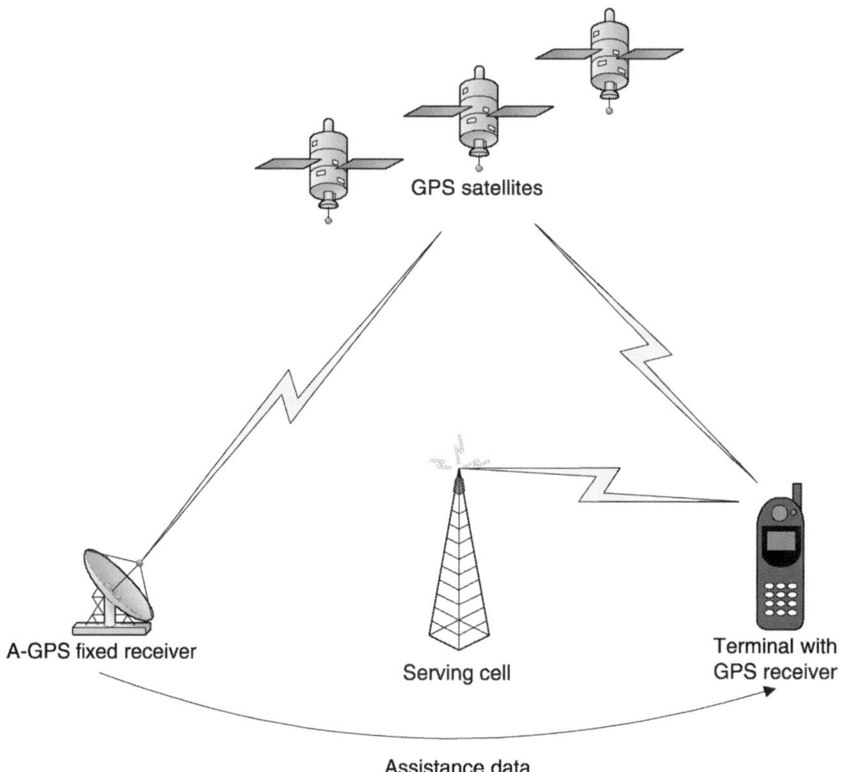

Figure 4.5 A-GPS delivery of assistance data.

The most basic and inaccurate method for locating mobile terminals is to use the cell identification (cell-ID) of their serving cell. The cell-ID typically defines a particular sector of the cell site, where a site can have up to six sectors that cover the surrounding circular area. To determine a user's serving cell, the network pages the user's terminal, or tracks the terminal's cell updates whenever it moves between locations. Since cell size radii can vary from about 150 meters in urban areas to 30 kilometers in rural areas, this method does not provide consistent accuracy.

The network-based OTDOA method ([10], [13]), also referred to as Uplink Time Difference Of Arrival (U-TDOA), uses Location Measurement Units (LMUs) installed at a network operator's cell sites to record the time of arrival of radio signals sent by the mobile terminals (see Figure 4.6). This approach is similar to the handset-based OTDOA method, except that the signal listening is done by the LMUs and computing is done at a central site, the Location Service Center (LSC), where measurements are collected from at least two distinct pairs of LMUs for each position computation. Some of U-TDOA's distinguishing features include the ability to compute positions that have an accuracy that is typically within 50 meters, provided there are sufficient tracking cell sites in the user's vicinity.

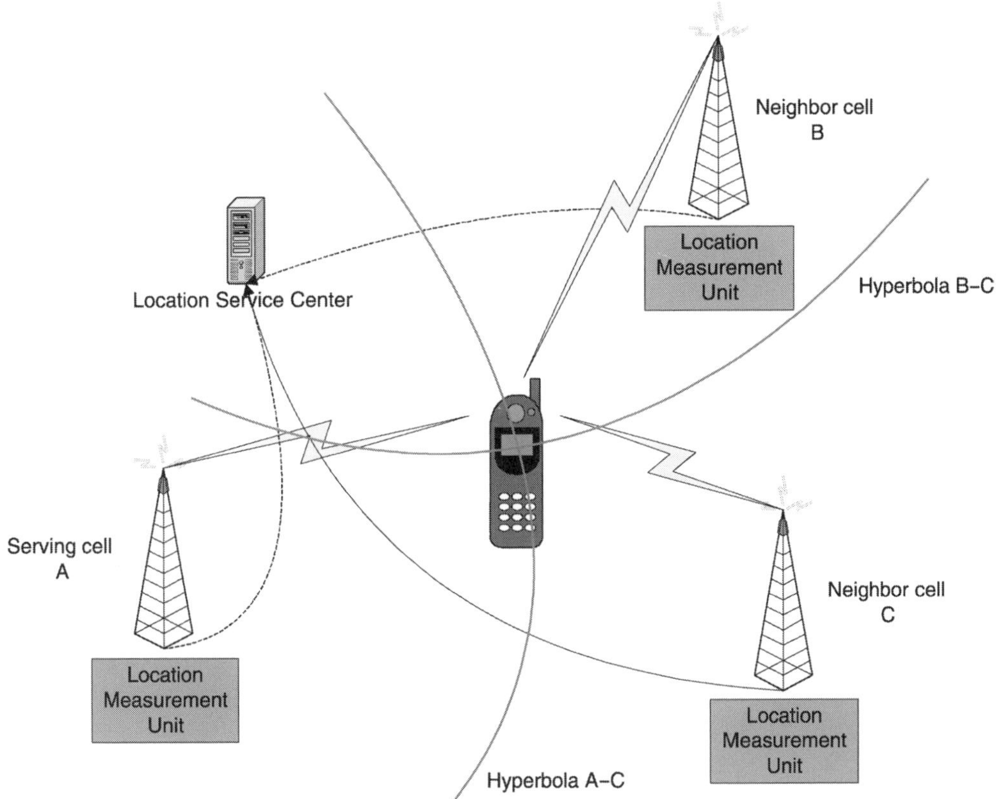

Figure 4.6 Network-based OTDOA position determination.

4.3 Location data access

Standard bodies have addressed the need for a common approach to access user location information that is independent of the particular underlying network technology. The 3GPP standards organization has formulated the Open Service Access (OSA) specification that allows third party mobile applications to access network services, including the access to location information [21]. Likewise, the Location Working Group (WG) of the Open Mobile Alliance (OMA) continues the work initiated by the former Location Interoperability Forum (LIF) and the former WAP Forum and developed specifications to ensure interoperability of mobile location services on an end-to-end basis [22]. The Location WG specified the Mobile Location Protocol (MLP), an application-level protocol for getting the position of mobile terminals independent of underlying network technology [23]. MLP is a request/response protocol that serves as the interface between a location server and a requesting client application, and can be mapped on top of HTTP.

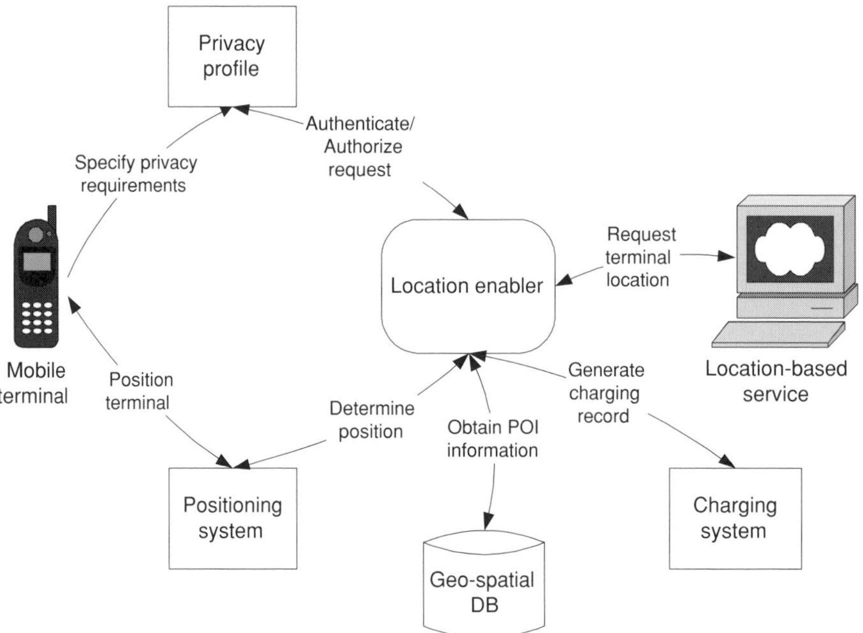

Figure 4.7 Context model for a location enabler.
(Source: OMA [24].)

The context model diagram in Figure 4.7 (adapted from [24]) illustrates a typical location request/response flow, with no implication for the placement of functions within specific physical entities. In this diagram, a location enabler function interacts with client applications: a client MLP request is issued by a location-based service, and the MLP response is returned after obtaining the terminal's position from a positioning system. OMA's location specification defines the core set of operations that a location server that supports a location enabler function should be able to perform. These include:

1. **Standard location immediate service**. This is a standard service for requesting the location of one or more mobile users. The service is used when a location response is required immediately. It also has support for requesting a certain quality of service, that is, the desired accuracy of the location and the asked for response time.
2. **Emergency location immediate service**. The emergency location immediate service is used to retrieve the position of a mobile user that is involved in an emergency call.
3. **Standard location reporting service**. A standard location report is generated when a mobile user wants a client application to receive his or her location. The client application that the location report should be sent to is specified by the mobile user or defined within the location server.
4. **Emergency location reporting service**. An emergency location report is generated if the wireless network spontaneously initiates a position determination when a user

initiates or releases an emergency call. The applications that the emergency location report is sent to are defined within the location server.

5. **Triggered location reporting service**. The triggered location reporting service is used when an application wants the position of a list of mobile terminals to be tracked. The triggers could be an elapsed period of time, or a mobile terminal action, defined as establishment of contact between the network and the mobile terminal (the event "UE available" defined by 3GPP in [25]). The report will be sent to the requesting client application when one of the above triggers occurred.

An example [23] of the location information, encoded in XML, returned by the location server to a client that issued a *standard location immediate* request is shown in Listing 4.1.

```
<slia ver="3.0.0" >
 <pos>
  <msid>461011334411</msid>
  <pd>
   <time utc-off="+0200">20020623134453</time>
   <shape>
    <CircularArea srsName="www.epsg.org#4326">
     <coord>
      <X>30 16 28.308N</X>
      <Y>45 15 33.444E</Y>
     </coord>
     <radius>240</radius>
    </CircularArea>
   </shape>
  </pd>
 </pos>
 <pos>
  <msid>461018765712</msid>
  <poserr>
   <result resid="10">QOP NOT ATTAINABLE</result>
   <time>20020623134454</time>
  </poserr>
 </pos>
</slia>
```

Listing 4.1 Standard Location Immediate answer

For the first mobile with ID *461011334411*, as specified in the first *<msid>* element, the *<time>* element gives the timestamp (in the format YYYYMMDDHHMMSS) when the location was measured. The *<CircularArea>* element provides the nautical latitude and longitude coordinates of the user's location, and the *<radius>* element is a distance measurement (in meters) from the circle's center that expresses an uncertainty in the actual location as the user could be in any point of the specified circle. The location server could not provide a location indication for the second mobile with ID *461018765712*, and returns an error indication in the *<poserr>* element.

4.4 Location data models

Applications that leverage location information can be partitioned into two main categories: mobile resources management and consumer applications. Mobile resources management includes mobile workforce allocation and fleets' vehicle tracking. Dispatch centers can use location data to perform varied functions such as schedule service staff visits, inform customers of service staff arrival times, track security vehicles, and direct mobile emergency staff in disaster situations. On the other hand, consumer applications serve mobile subscribers' messaging needs, answer information content requests, or send unsolicited notifications. The wireless network has to be appraised of a user's location to be able to send a message there. Similarly, more pertinent information can be sent to users, for example, a list of preferred restaurants in their immediate vicinity, when their location is known. Other consumer applications include location-based games. Some mixed-reality games (for example, BotFighters [26]) use the physical position of players as the game's backdrop where, depending on the players positions, certain actions become possible such as "shooting" other players.

Moving users are tracked in location databases that keep a record of each user's current location, and potentially also track a history of user locations. Location databases are queried using "instantaneous queries", which are evaluated once, when user locations need to be determined for completing transactions such as setting up a call, finding team members of a given user, or locating services in a user's vicinity. Location data is special in that it is continuously changing, and has a temporal dimension that indicates the data's time validity. In fact, the user's movement behavior determines the period of time for which queried location data is valid. For example, for a relatively fast traveling user, the queried location data may become out of date sooner. In such cases, "continuous queries", which are queries that are re-evaluated often, can be applied to prevent situations where the user is presented obsolete data. What circumstances would warrant continuous queries? Tourists in a moving car may request a list of hotels or restaurants in their vicinity. Such a list would need to be continuously updated to be of value to the moving tourists. To complicate things further, location queries may refer to future time, for example, a tourist driving on a highway may ask for a list of hotels that he or she will reach two hours from now.

In the following, we describe a progression of location data models, from models that just track user location, as in cellular networks, to spatial models that correlate

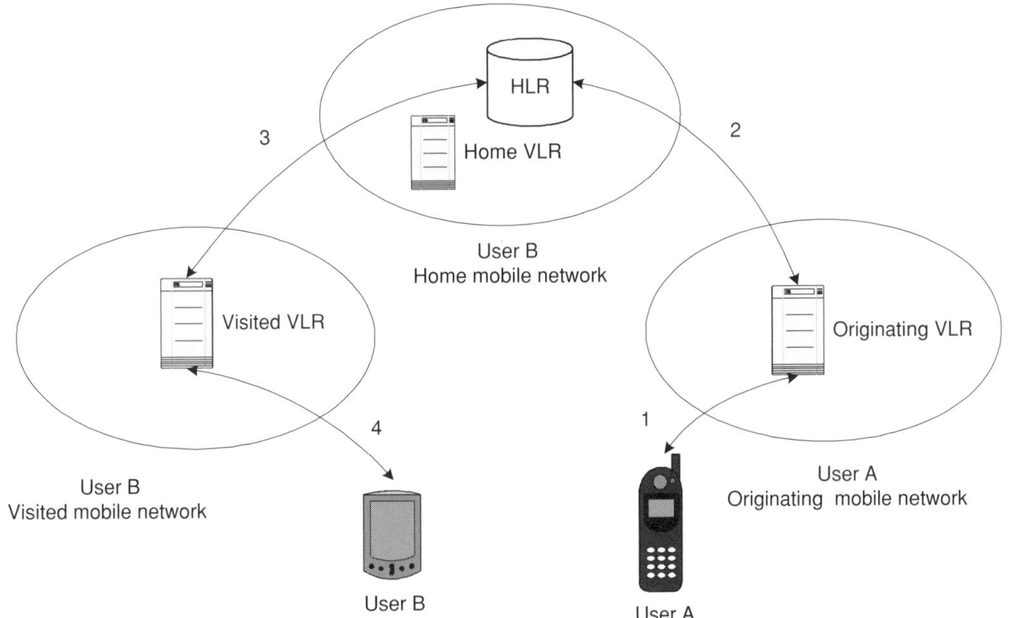

Figure 4.8 Two-tier storage scheme: tracking user locations in cellular networks.

between user location and nearby geographic data such as road segments, and finally spatial-temporal models that add a time-of-day aspect to location data.

4.4.1 User location data models

To efficiently track users, their locations need to be stored in network servers that can be queried when required. The alternative is not to track their location, and query the whole network, which is very costly in communication terms, and certainly not scalable for large numbers of mobile users. Unless the tracked population is relatively small, location data cannot be aggregated in one central location as this would lead to very expensive solutions for achieving good performance.

Cellular networks have adopted a two-tiered distributed approach for storing user location data [27] (see Figure 4.8). The top tier consists of a home location register (HLR) database. Every mobile subscriber is associated with one HLR that contains basic user information profiles that include an identification of the mobile area where they were last detected. With each mobile area is associated a visitor location register (VLR), a database of the bottom tier that contains copies of HLR entries for those mobile users that entered into the area administered by the VLR. When a message is sent or a call is issued to destination User B, the network first queries the originating VLR of User A, the message sender or caller (step 1 in Figure 4.8). If the destination user is not located in the originating VLR, then the destination user's HLR is queried

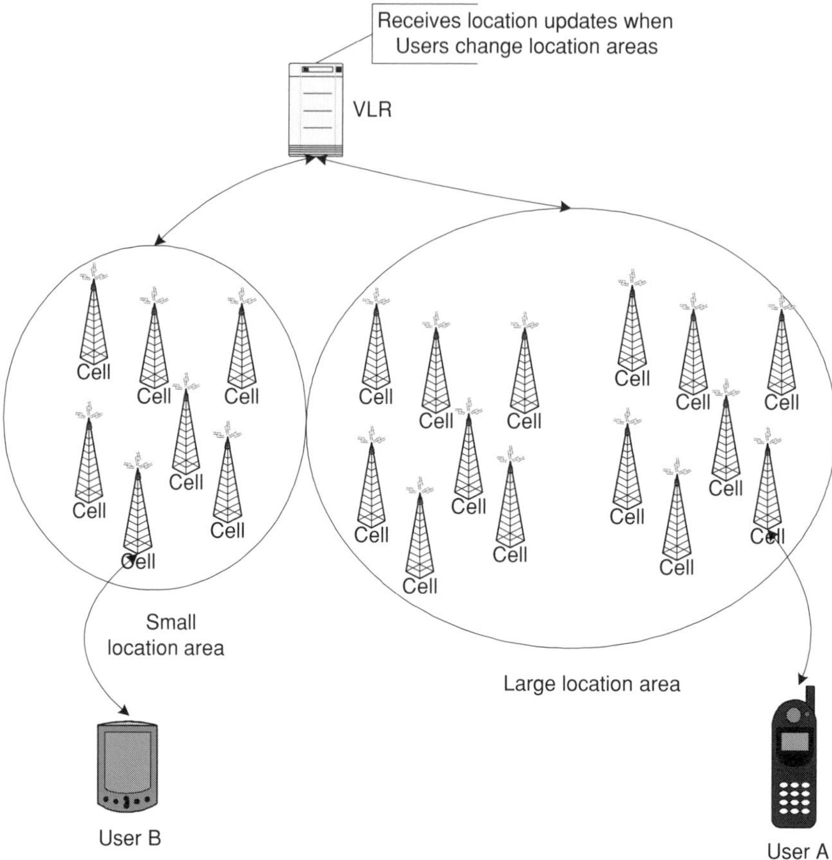

Figure 4.9 Cellular location areas for tracking users.

(step 2). When the destination VLR is found (step 3), a search query, referred to as paging, is issued to all cells in the VLR's area to identify the particular cell where the user can be found (step 4).

User tracking in cellular networks relies on location area-based algorithms that tend to reduce the cost of tracking user movements; however, this will occur at the expense of increased cost of setting up message communication or voice calls. Groups of cells form what are called "location areas" (see Figure 4.9) and a mobile user's terminal updates the user's location in the VLR when the user crosses boundaries between location areas. To find a user, all cells in a given location area are queried. If the location areas are large and include many cells, there will be fewer location updates so that the cost of tracking user moves is relatively low. However, the cost of establishing a call increases as more cells need to be queried to find a called user, incurring a larger communication overhead.

Multiple-level hierarchical schemes extend the two-tier scheme into a tree that includes additional levels with location databases at each node of the tree (Figure 4.10). A user's location, that is the cell where the user can be found, is specified in the database of one of the tree's leaf nodes. Internal nodes may contain pointers to lower-level

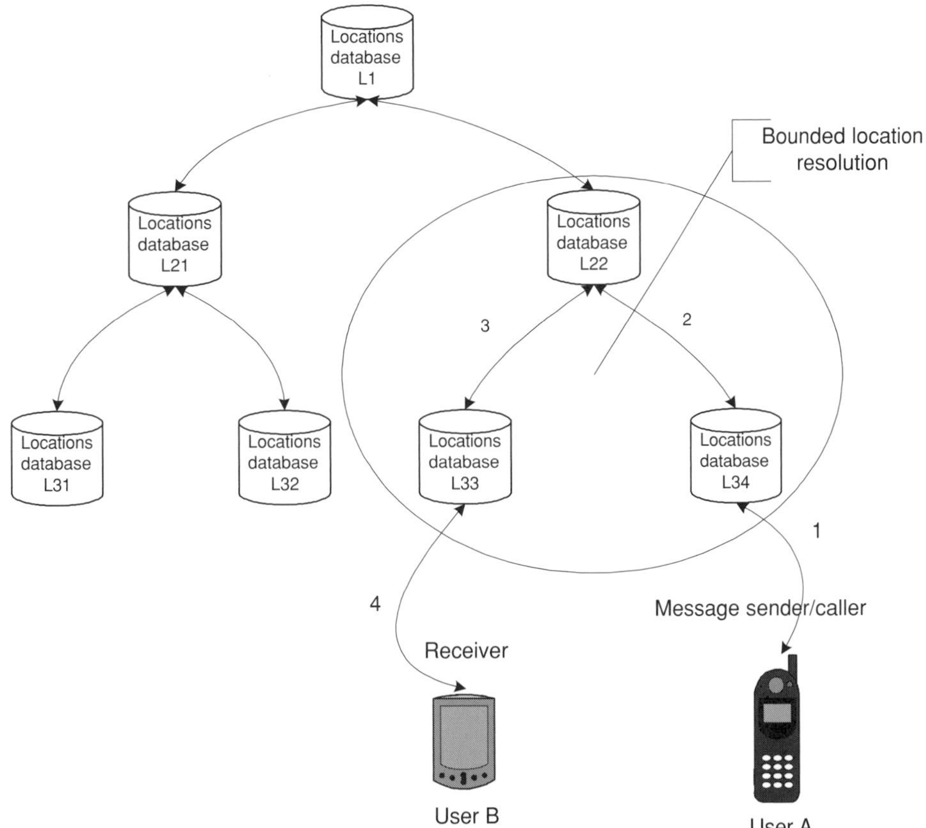

Figure 4.10 Hierarchical storage scheme: tracking user locations at multiple levels.

databases that contain the mobile user's location. Alternatively, each node's database can contain the location of all the users in the sub-tree below it. When a message is sent from User A at node x to User B at node y, then the search for User B will proceed from node x upward in the tree until a node database is found that contains User B's actual location or a pointer to User B's database (steps 1–4 in Figure 4.10). The hierarchical scheme has better support for message exchange or calls in a local area since there is no need to query a remote HLR for those users that are roaming into the current location area. However, content applications that wish to push personalized content to specific mobile subscribers need access to a centralized store that contains the mobile users' cell associations. These applications will have to exercise a search from the root of the tree, not unlike the case of the two-tier scheme.

With the above hierarchical storage structures are associated location management algorithms that specify where in the hierarchy to place location data, and what data replication and caching methods should be used at each level of the tree. The performance of a location management algorithm is determined by the relative frequency of user calls and moves as these affect the relative cost of lookups and updates of user locations (see [28] for a survey of location algorithms). For example, in a cellular two-tier scheme, user service parameters can be replicated at the VLR to reduce the number of lookups in the HLR at call setup time. Similarly, location information about called User B can be replicated in the VLR of an originating caller, User A, so that the same information can be reused in any subsequent calls to User B.

4.4.2 Spatial data models

Spatial databases are databases that store representations of spatial objects such as the space that embeds mobile user subscribers. These databases enable the correlation of user position with geographic data. For example, a spatial database could represent a city's road network, and the mobile subscribers would be assigned to road segments as determined by their current position. Another possible representation is to use a city's regions as specified in some geographic information system (GIS), and assign each user to a distinct region as long as there is no overlap between adjacent regions. With each space representation is associated a spatial indexing method that consists of "spatial keys" used for sorting purposes.

While mobile users are not part of the geographic spatial data, each mobile user record could include a spatial key that indicates current location. In addition, the spatial database can associate with each spatial object the non-geographic data that consists of mobile users that are currently within the spatial object's confines. By relating these two types of data, the location-based queries and proximity queries that could be answered include requests such as "find the location of user *m*", "find all users in a region *s*", and "find all users that are within *x* miles from landmark *l*".

Typically, hierarchical data structures are used to represent geographic spaces [29]. These data structures recursively decompose the space into regions also referred to as blocks. One approach would be to use what are termed minimum bounding rectangles (MBRs) to enclose the spatial objects of concern. For example rectangles could be used to enclose road segments. Each road segment would be associated with only one rectangle, although these rectangles may overlap.

R-tree, model of spatial objects

R-trees decompose the geographic data space, for example road segments, without regard to line or region coordinates. The R-tree data structure uses the MBR concept where each node of the tree corresponds to the smallest rectangle that encloses its children nodes, and rectangles of adjacent nodes may overlap. A variant of the R-tree, referred to as the R^+-tree, partitions space into non-overlapping rectangles. In this case, road segments may be associated with multiple rectangles, that is tree nodes, so that multiple nodes may need to be retrieved to determine the region covered by a spatial object.

An example R-tree with overlapping MBRs is shown in Figure 4.11, and the corresponding variant R^+-tree with no overlapping MBRs is shown in Figure 4.12. Both trees can help answer a proximity query such as "what are the roads that are near a

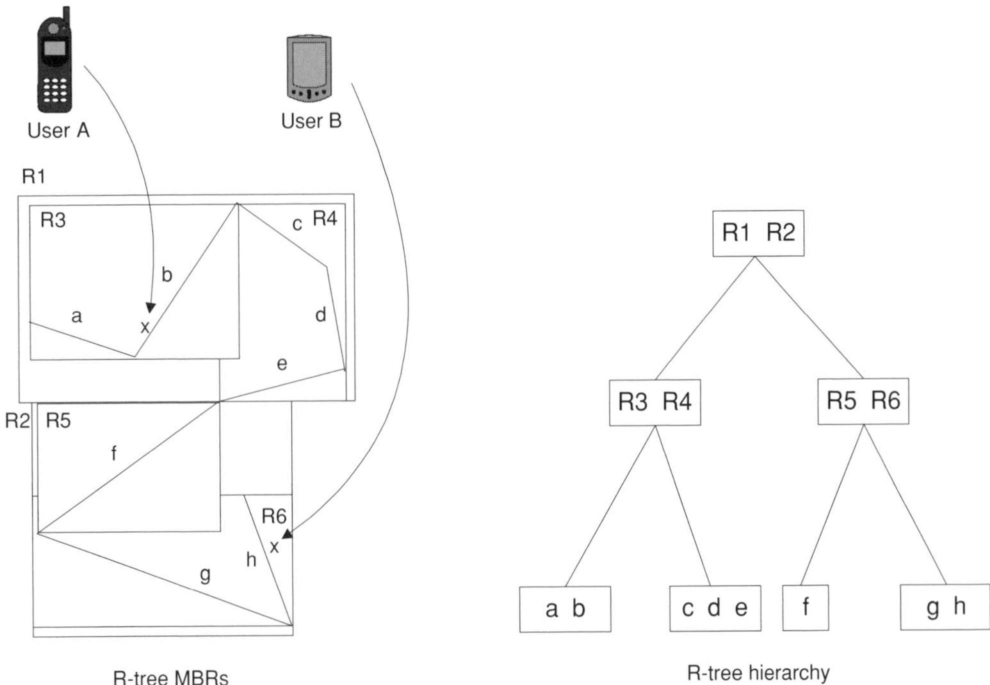

Figure 4.11 Road segments R-tree.

mobile user?" The user position coordinates determine in which MBR the user is situated. For example, a search through the R-tree in Figure 4.11 can find the neighboring road segments of User A; these are the segments within the same MBR where User A is located. These are the children of node *R3* and include road segments *a* and *b*. An extension of the search would look for the children of neighbor node *R4*, and come up with the additional road segments *c*, *d*, and *e*. Similarly, the R^+-tree in Figure 4.12 can help find the neighboring road segments of User A. These are the children of nodes *R3* and *R4*; however, in this case the duplicate road segment *b* should be discarded. Other proximity type queries include finding distances between mobile users. For example, a query can request the distance, expressed by the identified road segments, between User A and User B in the same figures.

Quadtree models of spatial objects

Both the R-tree and the R^+-tree use a space decomposition method that is dependent on the particular spatial objects, for example, road segments. Other recursive decomposition methods, referred to as Quadtree methods, are less data dependent and partition a geographic embedding space into a grid structure. The grid can consist of blocks of uniform size or else of sizes that adapt to the distribution of the data. Strictly speaking, the Quadtree is defined to have blocks with widths that are powers of two. Typically, a block will be decomposed when the application requires finer resolutions. For

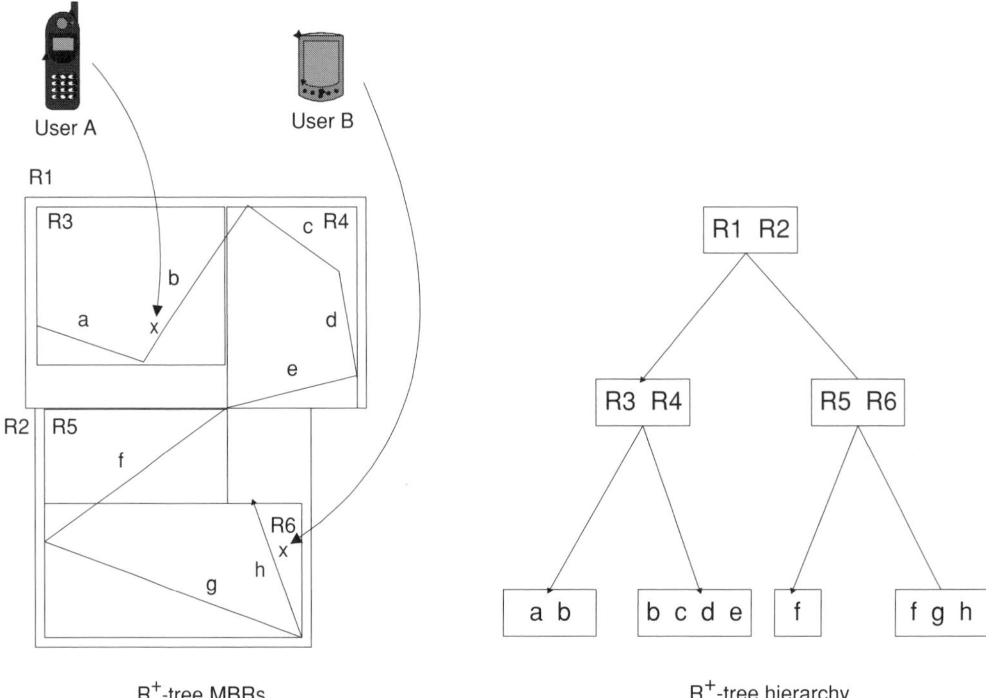

Figure 4.12 Road segments R^+-tree.

example, in the case of point data, a block can be decomposed when it contains more than a predetermined number of points of interest (POIs). This partitioning approach can restrict the number of point elements allowed in each block.

For line data, a similar decomposition can take place where, for example, a node of the Quadtree is restricted to contain no more than one line, unless the lines meet at the vertex contained in the node (see Figure 4.13). As in the R-tree case, location-based queries and proximity queries can be efficiently answered by traversing the Quadtree representation. The accuracy of query responses is improved as, for example, road *f* is determined to be far from User B, unlike the answer provided by the previous R^+-tree.

When mobile user positions need to be tracked with a fixed resolution, user movement can be tracked efficiently with a Quadtree scheme [30]. User position at each point of time is specified by the particular Quadtree block where the user happens to be. The required frequency of location updates is determined by the block size, with smaller sizes requiring more frequent updates. For example, blocks can have a side length of 3 miles if the grid covers an area where the average speed of a moving vehicle is no faster than 30 miles per hour (see Figure 4.14). Average block traversal time is then 6 minutes. A mobile user's GPS enabled terminal can have an update frequency of one location update every 3 minutes to determine the user's Quadtree block index. As

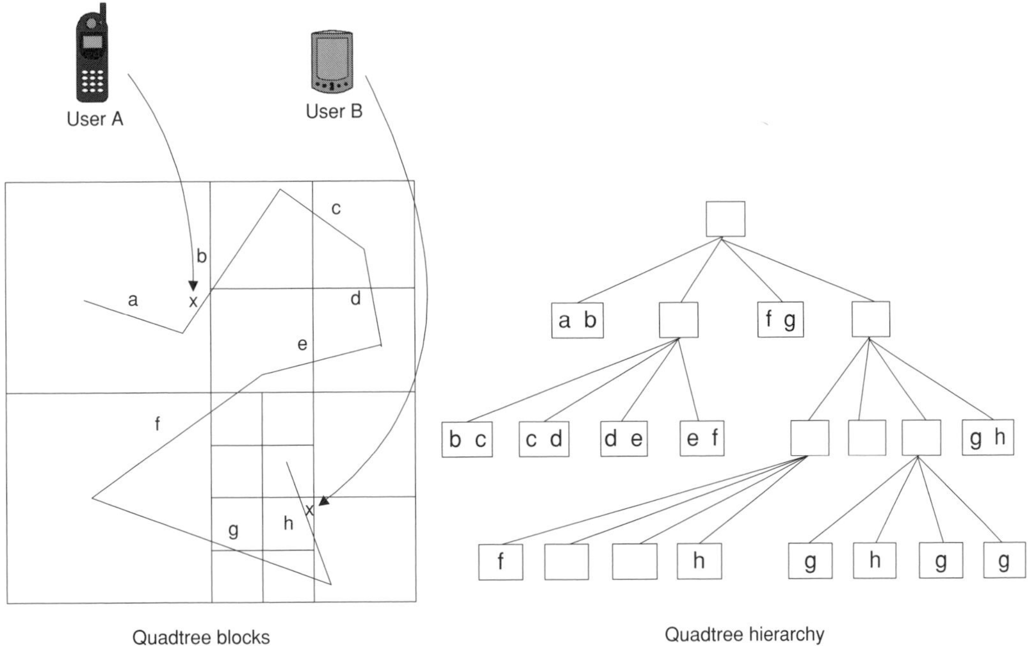

Figure 4.13 Road segments quadtree.

a result, at any point in time the user's location is within an uncertainty region with borders 1.5 miles away from the block from which the last location update was issued. For User A in Figure 4.14, this uncertainty region is the shown enclosure if a location update was last issued from the depicted user's position.

4.4.3 Spatial-temporal data models

In recent years, a number of research efforts took place to design special purpose location databases referred to as moving objects databases (MODs). These databases address the unique modeling requirements of location data as well as the indexing of moving objects such as mobile users, or a fleet's cars. MODs were designed to deal with deficiencies in standard database technology to efficiently handle continuously changing data, and the lack of modeling capabilities to represent queries that can simultaneously handle space and time constraints, including projections in the future. An example of such a query would be "retrieve the fleet's vehicles that will be in the town of Northbrook between 3 p.m. and 5 p.m.".

MODs have resorted to using dynamic attributes, that is, attributes that change dynamically over time without any update operations. For example, the position of an airplane or a car can be described by a function whose value changes with time depending on the route and speed of the moving object. This provides for a potentially

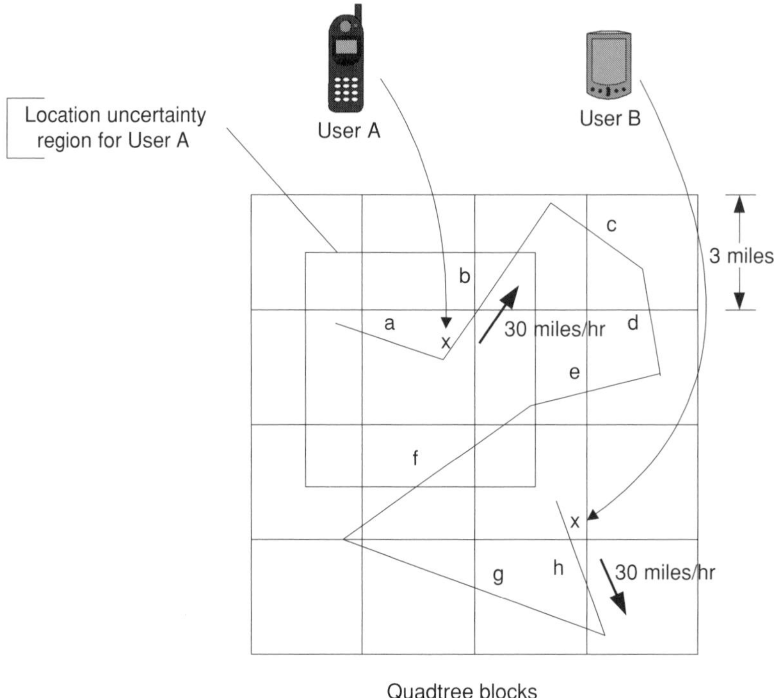

Figure 4.14 Road segments quadtree with location uncertainty region.

significant performance improvement over a traditional approach of frequent database location updates.

In addition to the expected performance improvement, modeling the position of moving objects as functions of time enables tentative future predictions. For example, it may be 5 p.m., and the traveling user could issue a query: "Give me the list of hotels no further than 5 miles from my location between 8 p.m. and 9 p.m.". Answers to these queries require the database to keep a model of user trajectories, and of events, referred to as triggers, that may cause a re-evaluation of the query so that the presented data remains accurate. For example, a change in the user's speed or direction may trigger a re-evaluation. There could also be external events such as traffic congestion, or accidents, which will impact the user's route or travel time. Continuous querying will ensure that the retrieved data is accurate, and a number of research efforts have investigated efficient schemes for query re-evaluations.

Various indexing techniques have been proposed to support the querying of current and future positions. Saltenis *et al.* [31], for example, proposed using R-tree-based indexing for retrieving moving objects within specified regions of moving objects. This technique partitions the moving objects into bounding rectangles that continuously follow the moving objects.

In addition to user location attributes in the form of trajectories that include a time dimension, other attributes need to be tracked in a MOD. Trajcevski and Scheuermann

[32] proposed keeping for each mobile user a list of the posed queries and a list of the queries that are dependent on this user. As part of the specification of the MOD, the designer specifies the "triggers" that will cause query re-evaluations, so that these re-evaluations are performed in an optimal fashion that does not cause re-evaluations at successive clock ticks. For example, a trigger could specify that a change in a user's speed or direction requires query re-computing. In the event of such a trajectory change, the MOD system will check if the user posed any queries, and if any queries posed by other users are dependent on this specific user, so that the affected queries can be re-computed. For example, for the query: "Give me the list of my co-workers that will be within 1 mile from my location at 3 p.m.", the MOD system will take into account the relationship between this query and the co-workers' identities to decide when to re-evaluate the query should there be any changes in the co-workers' trajectories.

As there are numerous mobile applications that could leverage MODs, practitioners have turned their attention to implementations on top of existing commercial databases. New special-purpose operators that pertain to spatial objects (points or regions) and trajectories were designed to manage simultaneously space and time information. For example, Varzirgiannis and Wolfson have defined a set of data types and query predicates in terms of SQL extensions [33] showing how to implement an SQL query that will retrieve moving objects that are located within a fixed travel time from a point on their route. Wolfson has provided a classification of the new operators [34]. These are partitioned into three classes: operators that query a single trajectory, operators that query the relationships between trajectories and spatial objects, and operators that query the relationships between multiple trajectories. The following queries are examples of each query type:

- Single trajectory query: "When will user A be closest to address X?"
- Relationship between spatial objects and trajectory query: "When will vehicle V be within 3 miles to a warehouse in region R between 1 p.m. and 5 p.m.?"
- Multiple trajectories query: "When will aircraft A be within 10 miles of aircraft B between 8 a.m. and 9 a.m.?"

4.5 Enhanced 911 service

Government efforts in both the USA and Europe have addressed the need to deploy wireless location detection solutions that can enable security staff to respond faster to emergency calls from wireless users in distress. Location knowledge can serve at least three aims [35]:

- Route the wireless emergency calls to the correct emergency call centre.
- Locate the caller and incident site.
- Dispatch the most appropriate emergency response team(s).

While the first two objectives are the responsibility of the wireless operator, the third one is dependent on the emergency service provider. This section focuses mainly on the objective to locate the caller.

4.5.1 Government mandates

Emergency call statistics show why it is imperative to provide adequate emergency service to wireless callers. In the US, it is estimated that of the 150 million calls that were made to 911 in 2000, wireless telephone users made 45 million of them [36]. This is a ten-fold increase since 1990, and it is anticipated that by 2005, the majority of 911 calls will be from wireless callers. In the European Union countries, about 50 percent of emergency calls (about 40 million calls per year) emanate from mobile phones [35]. About 15 percent of these calls provided inaccurate location or no location information, hampering the dispatch of emergency teams. For the remainder of the calls, significant time, up to a few minutes, could be saved if location information is automatically provided.

Planning for location-based emergency services in the USA started in 1996, when the US Federal Communications Commission (FCC) passed an Enhanced 911 (E911) mandate for the US nationwide 911 emergency service number [37]. The objective of the Phase I mandate was to provide emergency operators receiving a wireless call with the same information available to them when the emergency call is made from a wireline phone. Wireless carriers were required to provide to emergency service dispatchers, referred to as Public Safety Answering Point (PSAP), the serving cell site and telephone number of the originator of a 911 call. At the time when the mandate was issued, the available location technology was network-based. With the advancement in wireless technology, this rule was revised in 1999 to a Phase II mandate, to better enable carriers to use handset-based location technologies. This latest revision requires wireless carriers using a handset-based solution to be able to locate 67 percent of 911 callers within 50 meters, and 95 percent of callers within 150 meters. For a network-based solution the requirement is relaxed to 100 meters accuracy for 67 percent of the callers, and 300 meters accuracy for 95 percent of the callers. The caller's location coordinates include latitude, longitude, and altitude.

The European equivalent of the E911 service is E112 with a common 112 emergency number [35]. As of July 2003, European wireless carriers are required to deliver location information with each 112 call. However, there has been no mandate on the accuracy of the location, so most European wireless carriers use the serving cell's identification as the user's position. This accuracy varies with the cell size, as across a network cell sizes vary considerably. Larger cells are referred to as macro-cells and are typically a few tens of kilometers radius in rural areas. In suburban areas macro-cells are a few kilometers radius. In densely populated areas, micro-cells are often deployed and these cells are can range from 100 meters to 500 meters in radius. Finally, pico-cells can be deployed in buildings and these cells are a few tens of meters in radius.

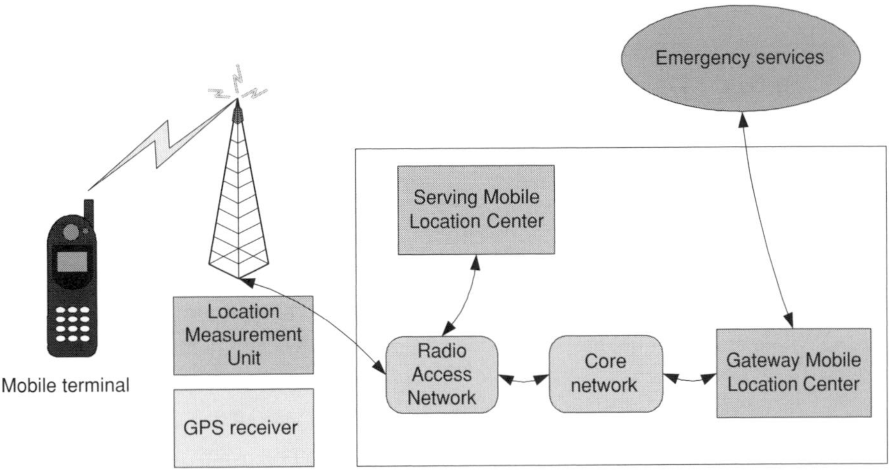

Figure 4.15 Location services architecture.

Macro-cells will frequently have three sectors, each pointing in a different direction. This enables the position of the mobile to be estimated more accurately than from an omni-directional cell. Furthermore, a parameter called the Timing Advance (TA) is used in normal GSM operation and can be used to improve location accuracy. TA is a measure of the range of the connected mobile from the cell site and is accurate to approximately 550 meters.

4.5.2 Location services architecture

The network architecture for location services adds a number of new network elements (see Figure 4.15). At the cell site, a unit called the location measurement unit (LMU) provides measurement data for network-based location technologies such as OTDOA. Alternatively, if a handset-based location technology such as A-GPS is used, then the cell site will contain a GPS receiver to assist in location computations. The serving mobile location center (SMLC) specified by 3GPP in [38] is responsible for the collection and coordination of all the necessary measurements in the network, including in the mobile terminals, to give a position estimate of the mobile subscriber. Finally, the Gateway Mobile Location center (GMLC) manages the external interface to outside location-based applications, including emergency service providers. It provides an interface for retrieving mobile user location information, and performs authorization and privacy checks as well as provisioning and billing.

As of October 2003, the National Emergency Number Association noted that just 10 percent of 6000 US emergency call centers have the ability to locate the precise site of wireless callers calling 911. US cellular operators have until December 2005 to comply with the E911 ruling when 95 percent of all mobile terminals in service will have to support accurate location detection. Rollout of E911 has begun, with, for example, Sprint PCS offering this capability since the end of 2001.

For GSM Edge networks, the 3GPP standards organization has established three positioning methods [39]. These include the Enhanced Observed Time Difference of arrival (E-OTD) which is similar to the previously described handset-based OTDOA, the GPS and A-GPS methods, and the Uplink Time Difference Of Arrival (U-TDOA) which is similar to the network-based OTDOA. In the U-TDOA method each cell site is equipped with an LMU to monitor the transmission duration time of signals sent by mobile terminals to cell sites (see the similar OTDOA architecture in Figure 4.6). The LMU measurements are forwarded to a central SMLC where the terminal's position is computed. The terminal's position and a confidence region associated with the location estimate are computed in a similar way as in OTDOA. The measurements, processing, and reporting take no more than a few seconds. This time is short enough to allow a call to be routed to a service provider that is close to the user's position. As this method requires at least three cell sites, it provides accurate results in areas with a high density of cell sites, such as urban areas; however, it is less accurate in rural environments where only one cell site may be available. The US GSM operators, Cingular Wireless and T-Mobile, for example, support U-TDOA for determining mobile user locations [40], [41].

Cellular networks that use code-division multiple-access (CDMA) for wireless communications have relied on a handset-based location determination approach based on the A-GPS method (see Figure 4.5). The wireless carrier's network provides indications to the mobile terminals that suggest, based on the location of the current connected-to cell, the best GPS satellites to listen to for location computation. These indications are particularly helpful when reception is not very good, for example when the mobile user is indoors, although this information is of no value when the mobile terminal is deep inside a building and cannot receive any satellite signal. Deploying an A-GPS-based system would typically require each cell site to include an added GPS receiver that can provide the list of satellites that mobile users in its area should listen to. The US operators Sprint PCS and Verizon Wireless are examples of CDMA wireless operators that support an A-GPS-based solution to fulfill the FCC's E911 requirements [42].

The US government rules and mandates for wireless emergency procedures have, as seen above, led to the development of location detection technologies. Advances were made in both the network and the terminals with new solutions that provide relatively accurate positioning. These same technologies will be used to develop the market for location-based services targeted at the mobile consumer. Besides the basic benefit of presenting the user with current location data, for example in the form of a dot on a map, the added benefit of automated position determination is ease of use when interacting with services that leverage this data. Furthermore, location-aware services could also send the user unsolicited information that he or she may want to know about when moving into a new area.

REFERENCES AND FURTHER READING

[1] S. Deering and R. Hinden, *Internet protocol, version 6 (IPv6) specification*, RFC 2460. IETF (Dec. 1998).

[2] R. Droms, *Dynamic host configuration protocol*, RFC 2131. IETF (Mar. 1997).

[3] Nokia, *Transition to IPv6 in 2G and 3G Mobile Networks*, White paper (2000).

[4] C. Perkins, *IP Mobility Support*, RFC 2002. IETF (Oct. 1996).

[5] J. D. Solomon, *Mobile IP: The Internet Unplugged*. (New Jersey: Prentice Hall, 1998).

[6] S. Thomson and T. Narten, *IPv6 Stateless Address Autoconfiguration*, RFC 2462. IETF (Dec. 1998).

[7] Greenwich 2000 Limited, http://www.greenwich2000.com.

[8] Greenwich Guide, http://www.greenwich-guide.org.uk/meridian.htm.

[9] US Census Bureau, *Map Surfer*, http://www.census.gov/geo/www/tiger/tigermap.html.

[10] J. J. Caffery and G. L. Stuber, Overview of radiolocation in CDMA cellular systems. *IEEE Communications Magazine* (Apr. 1998), 38–45.

[11] 3GPP, *Stage 2 Functional Specification of User Equipment (UE) Positioning in UTRAN*, TS 25,305. (Sep. 2003).

[12] Y. Zhao, Standardization of mobile phone positioning for 3G systems. *IEEE Communications Magazine* (Jul. 2002), 108–16.

[13] C. Drane *et al.*, Positioning GSM telephones. *IEEE Communications Magazine* (Apr. 1998), 46–59.

[14] The Aerospace Corporation, *GPS Primer*. (http://www.aero.org/publications/GPSPRIMER/, Jul. 1999).

[15] R. Bajaj *et al.*, GPS: Location tracking technology. *IEEE Computer* (Apr. 2002), 92–4.

[16] J. G. McNeff, The global positioning system. *IEEE Trans. on Microwave Theory and Techniques*, **50**:3 (Mar. 2002), 645–52.

[17] The European Space Agency, *What is Galileo?* http://www.esa.int/esaNA/GGGMX650NDC_index_0.html.

[18] G. M. Djuknic and R. E. Richton, Geolocation and assisted GPS. *IEEE Computer* (Feb. 2001), 123–5.

[19] Cellular-news, *DoCoMo to start selling GPS handset*. (http://www.cellularnews.com/story/8706.shtml, Apr. 17, 2003).

[20] Motorola, *i730 Phone User's Guide (2003)*, http://idenphones.motorola.com/iden/iden_home.jsp.

[21] 3GPP, *Open Service Access (OSA) API, Part 1: Overview*, TS 29.198-1. (Dec. 2003).

[22] H. Dieterich and S. Deshmukh, *High-level requirements*. OMA Location WG. (http://member.openmobilealliance.org/ftp/public_documents/loc/2003/OMA-LOC-2003-0133R02-High_Level_Requirements_for_ArchOverview.zip, Oct. 2003).

[23] OMA, *Mobile location protocol 3.2.0*. OMA Location WG, OMA-LIF-MLP-V3.2.0-20031114-D, Nov. 2003.

[24] S. Deshmukh and A. Mattila, *Architecture overview and context model in architecture document*. OMA Location WG. (http://member.openmobilealliance.org/ftp/public_documents/loc/2003/OMA-LOC-2003-0279-LATE-ArchOverview_Domain_and_Context_Model_in_AD.zip, Dec. 2003).

[25] 3GPP, *Functional Stage 2 Description of LCS*, TS 23.271. (Dec. 2003).

[26] BotFighters, *It's Alive*. http://www.botfighters.com.

[27] M. Mouly and M. B. Pautet, *The GSM System for Mobile Communications*. (Telecom Publishing, 1992).

[28] E. Pitoura and G. Samaras, Locating objects in mobile computing. *IEEE Trans. on Knowledge Engineering*, **13**:4 (2001).

[29] H. Samet, *The Design and Analysis of Spatial Data Structures*. (Reading, MA: Addison-Wesley, 1990).

[30] A. Pashtan *et al.*, CATIS: A Context-Aware Tourist Information System. *Proc. of the 4th International Workshop of Mobile Computing*, Rostock, June 2003.

[31] S. Saltenis *et al.*, Indexing the positions of continuously moving objects. *SIGMOD* (2000).

[32] G. Trajcevski and P. Scheuermann, Triggers and continuous queries in moving objects databases. *6th Int. Workshop on Mobility in Databases and Distributed Systems (DEXA 2003)*, Sep. 2003.

[33] M. Vazirgiannis and O. Wolfson, A spatio-temporal model and language for moving objects on road networks. *7th Int. Symp. on Spatial and Temporal Databases*, Jul. 2001.

[34] O. Wolfson, Moving objects information management: The database challenge. *Proc. of the 5th Workshop on Next Generation Information Technologies and Systems (NGITS 2002)*, Jun. 25–6, 2002.

[35] European Commission, Coordination group on access to location information for emergency services (CGALIES), DG INFSO. (http://europa.eu.int/comm/environment/civil/pdfdocs/cgaliesfinalreportv1_0.pdf, Feb. 2002).

[36] National Emergency Number Association (NENA), http://www.nena.org/Wireless911/.

[37] Federal Communications Commission, Enhanced 911. (http://www.fss.gov/911/enhanced/).

[38] 3GPP, *Mobile Station (MS) – Serving Mobile Location Centre (SMLC) Radio Resource LCS Protocol (RRLP)*, TS 44.031. (Feb. 2004).

[39] 3GPP, *Functional Stage 2 Description of Location Service (LCS) in GERAN*, TS 43.059. (Nov. 2003).

[40] M. Rockwell, *Cingular switches E911 strategy to U-TDOA*, http://www.itsa.org/ITSNEWS.NSF/0/c33e6478b0ffe49585256c93004ac7dc?OpenDocument.

[41] TruePosition Inc., *Trueposition Press Releases*, http://trueposition.com/news_pressrelease.html.

[42] Federal Communications Commission, *Fact sheet: E911 phase II decisions*. (http://www.fcc.gov/Bureaus/Wireless/News_Releases/2001/nw10127a.pdf, Oct. 2001).

3GPP, *Broadcast Network Assistance for Enhanced Observed Time Difference (EOTD) and Global Positioning System (GPS) Positioning Methods*, TS 44.035. (Dec. 2002).

S. Prabhakar *et al.*, Query indexing and velocity constrained indexing: Scalable techniques for continuous queries on moving objects. *IEEE Trans. Computers*, **51**:10 (2002).

S. Saltenis and C. Jensen, Indexing of moving objects for location-based services. *Int. Conf. on Data Eng.*, 2002.

H. Samet, Spatial data structures. In *Modern Database Systems: The Object Model, Interoperability, and Beyond*, ed. W. Kim. (Addison-Wesley/ACM Press, 1995), pp. 361–85.

H. Samet and W. G. Aref, Spatial data models and query processing. In *Modern Database Systems: The Object Model, Interoperability, and Beyond*, ed. W. Kim. (Addison-Wesley/ACM Press, 1995), pp. 338–60.

U.S. Census, Bureau, *Topologically Integrated Geographic Encoding and Referencing System (TIGER)*, http://www.census.gov/geo/www/tiger/.

J. Widom, The Starbust active database rule system. *IEEE Trans. Know. and Data Eng.*, **8**:4 (1996).

O. Wolfson *et al.*, Updating and querying databases that track mobile units. *Distributed and Parallel Databases*, 7 (1999).

5 Wireless network security

Security of the connection between a mobile terminal and a network content server covers many aspects that are the focus of this chapter. We review the objectives of security and the assignment of responsibilities between concerned parties. Initially, we describe the commonly applied methods for securing the transmission of user requests and the returned content. We proceed next with a depiction of how the end-points of the communication link, that is, the terminal and the server, are authenticated, and lay out an authorization framework with the associated message sequences for granting access to content. As the mobile user will often want to access Web-based services, we describe next the standards developed by the OASIS standards organization for securing Web services. Finally, we conclude with a review of some of the industry de facto security methods for public-key cryptography.

5.1 Security objectives

Wireless network security addresses multiple concerns, both of the mobile subscriber and the Web information service provider. First, the user wants to verify that he is talking to the right server: this is "server authentication". In addition, the mobile user wants to be assured that the data transmitted over the air to a Web server remains "confidential"; in other words, it cannot be intercepted and understood by eavesdroppers. The user also wants to know that the data has not been tampered with en route, that is, "data integrity" is guaranteed.

The Web information service provider has other concerns, essentially "user authentication" and "authorization". Any mobile user requesting a service needs to be authenticated to make sure that he or she is not impersonating someone else (referred to as "spoofing"). Related to this is the concept of "non-repudiation" that guarantees that the sender of a request cannot deny his or her identity. Once authenticated, the user's data access authorization rights are set, he or she can access relevant data according to assigned access rights. These rights could have been set a priori, for example, according to a class of service that the user has subscribed to or to the role played by the user (Has the user paid for a higher level of service? Is the user acting in a group leader role where he has access to additional information?). Alternatively, access rights could be determined on the fly, for example, according to the user's willingness to pay extra for more information.

In addition to the security concerns listed above, wireless networks have further security issues with respect to over-the-air transmission and additional gateways in between the wireless and wired domains. In a wireless environment there are more possibilities for eavesdropping, and the additional gateways provide additional opportunities for tampering with the data in transit. Often, unsolicited information pushed to client mobile terminals needs to be accommodated too. Client and server roles are reversed in this latter case, with the client being the network infrastructure server that wishes to notify the mobile user of some new information. Before pushing its content, the server will want to ascertain that the recipient is in fact the intended mobile terminal. For example, a bank wishing to inform a mobile user about a financial transaction will first want to authenticate the mobile terminal before pushing this confidential information.

Various standards organizations have dealt with wireless networks security requirements. The security architecture in 3G cellular networks was formulated by the 3GPP standards organization in [1], [2], and [3], and CDMA networks security requirements were addressed in the IETF's RFC 3141 [4]. The mechanisms and procedures needed to implement wireless security are described in the following sections.

5.2 Security domains

Security requests and assessments often span multiple administrative domains. Typically there will be a user's home domain, for example the user's Internet Service Provider (ISP) or cellular network operator that provides the connectivity facilities, and a content provider's domain where the accessed wireless content is maintained. When mobile users roam outside of their home cellular network, they access a visited network. Bilateral agreements between different cellular network operators in the form of contracts, or service level agreements, will ensure that a user can access services outside of a home domain and be billed accordingly.

Typically, the content provider is a third party that delivers mobile content through the network operator's transmission capabilities, although in many cases the cellular network operator also plays the role of content provider. If the cellular operator and content provider are distinct, then each domain will each have its own authentication, authorization, and accounting (AAA) server. In some cases, the cellular operator will perform AAA operations on behalf of the content provider for a fixed fee charged to the content provider. This is, for example, NTT DoCoMo's mode of operation, as it charges the content providers 9 percent of the content access fee that is paid for by the mobile subscribers [5]. Mobile subscribers receive just one bill from DoCoMo that includes the charges for content subscription.

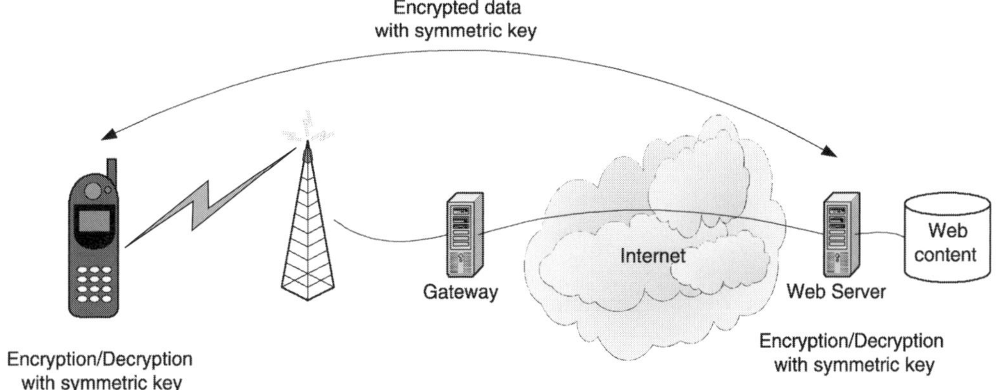

Figure 5.1 Symmetric-key encrypted data sent between mobile terminal and content provider.

5.3 Data transmission protection

To address transmission security concerns, the same security solutions applied in the wired environment can be used in wireless networks. Public-key cryptography standards are leveraged in the prevailing security solutions to ensure confidentiality of the transmitted data. Encryption algorithms, referred to as ciphers, include the RCx ciphers from RSA Security [6], the Data Encryption Standard (DES) from the National Institute of Standards and Technology (NIST) and variants of DES [7]. Encryption was also specified for XML in XML Encryption [8], a W3C specification for the selective encryption of XML documents where specific elements can be encrypted.

5.3.1 Symmetric and public-key cryptography

A cryptography algorithm, also referred to as a cipher, is applied to encrypt a message. At the receiving end, message decryption is performed. Typically, the ciphers are published and known; however, they use keys which are kept secret as decryption without knowledge of the used key would be extremely difficult, or even impossible. In "symmetric-key" encryption the same private key is used at both ends of the communication channel for both encryption and decryption. Critical to the operation is the secrecy of the shared key, so that authentication is also perceived as being met by the fact that no one else supposedly has knowledge of this key (Figure 5.1).

Diffie and Hellman introduced "Public-key" cryptography, also referred to as asymmetric cryptography, in 1976 [9]. The widely used RSA algorithm, invented in 1977 by Rivest, Shamir, and Adelman, supports a public-key infrastructure (PKI) [10]. In a public-key system, the receiver advertises a public key X that is used to encrypt data. Any sender has access to the public key and uses it to encrypt data. The receiver has a

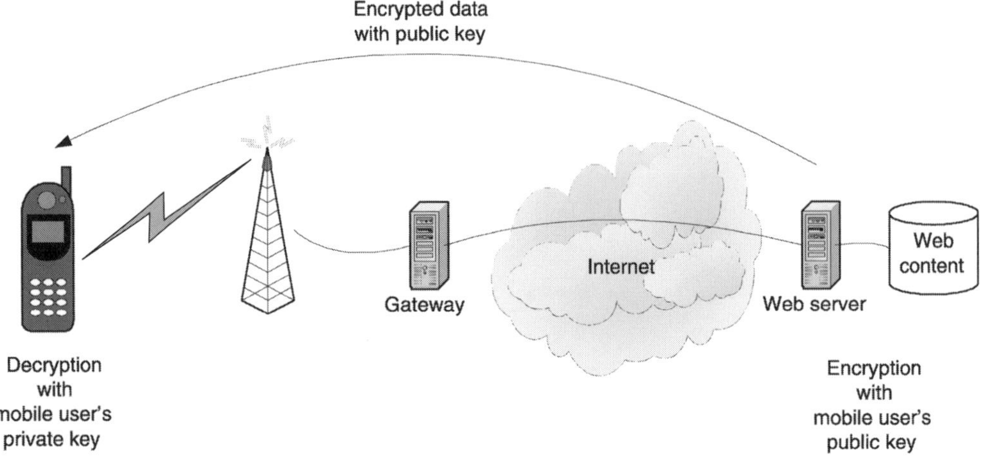

Figure 5.2 Public-key encrypted data sent to a mobile terminal.

private key K that only it knows about, and that is needed to decrypt the data. No other party can decrypt the messages encrypted with the associated public key. In Figure 5.2 the data sent by the content provider is encrypted with the mobile user's well-known public key. The users' public key could, for example, be sent by the mobile terminal to the content provider's site as part of a data request message that is not encrypted. Once the encrypted data is received by the mobile terminal, the received data can be decrypted with the user's private key.

Typically, public-key ciphers have a high overhead compared to symmetric-key ones. The reason for this is that public-key ciphers have to use longer keys, for example, 1024 bits for achieving the same encryption strength as symmetric-key encryptions since the former ciphers are limited to use a smaller set of keys because of the way they are computed. Public keys are typically generated by multiplying two large prime numbers. For example, banks typically require 128-bit keys for encryption in online banking. For this reason, it is more economical to send a "session key" that is encrypted with the public key. Upon receipt of this session key, the receiver decrypts it and uses it to decrypt sent messages, as is done in symmetric-key encryption. In other words, public-key encryption is used to transfer a private key that is used in turn to encrypt messages between communicating parties. Setting this shared private key is referred to as a handshake described in a following section on the SSL protocol.

The key length determines a cipher's encryption strength, that is, the difficulty of cracking a message encrypted with this key [11]. Although longer keys could be used, 128-bit keys are considered strong encryption, and security products that relied on such keys were once banned from export by the US government. Since January 2000 this restriction no longer holds.

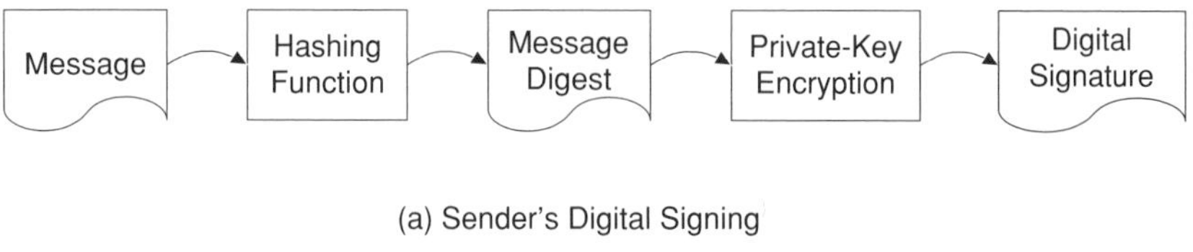

(a) Sender's Digital Signing

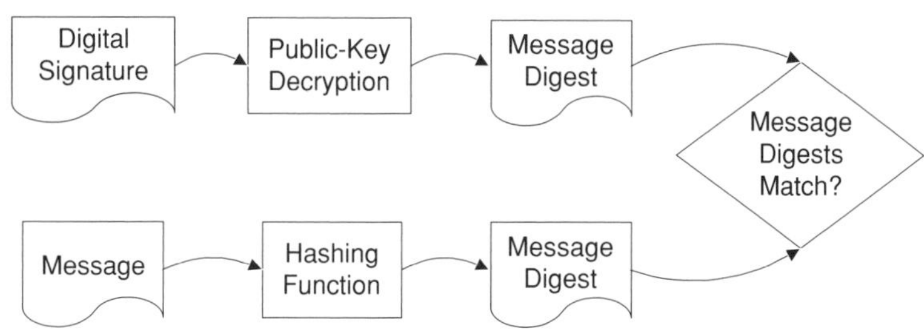

(b) Receiver's digital signature decryption and message integrity check

Figure 5.3 Message integrity verification with digital signatures.

5.3.2 Digital signatures

Data encryption addresses the data confidentiality concern, and digital signatures add another level of protection that addresses the user's concern for data integrity. Although digital signatures are also used to authenticate a user similarly to a physical signature, they are mainly used to ascertain that user data was not tampered with while in transit. The operation of digital signatures relies on the mechanisms of public-key cryptography. Unlike the use of a public key for data confidentiality, as described above, this time the sender uses a private key to encrypt a representation of the message (Figure 5.3).

First, for performance reasons, the sender uses a hashing algorithm, for example MD5 [12] or SHA-1 [13], to create a message digest that is a much shorter representation of the original message. The MD5 algorithm takes a message of arbitrary length and produces a 128-bit message digest, while the SHA-1 algorithm creates a 160-bit digest. This message digest is then encrypted with the sender's private key. The result is referred to as a "digital signature". Both the original message and the separate digital signature are sent to the receiver site. The original message could be encrypted too using, for example, symmetric key encryption, although this is not required for the digital signature's operation.

The receiver of the message first decrypts the digital signature using the sender's public key. The result is the sent message digest. Then, the receiver applies the same hashing algorithm on the original message to compute a message digest. If the decrypted and generated message digests are identical, then the receiver is assured that the data

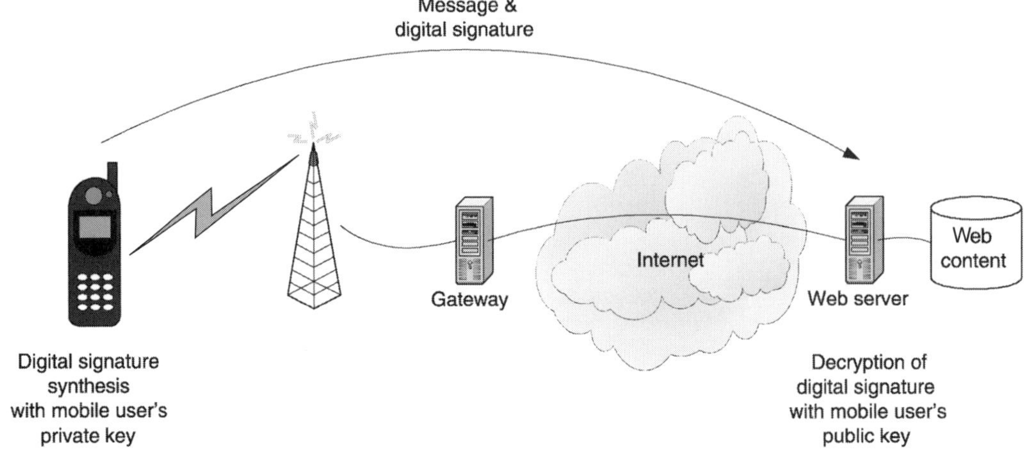

Figure 5.4 Digitally signed data sent by mobile terminal.

was not altered on the way. If they are not the same, then either the data was changed or some error occurred in transit. A mobile terminal sender and Web server receiver site operation are illustrated in Figure 5.4. Digital signatures also help ensure non-repudiation, where the sender cannot deny having sent the message, unless his private key was stolen.

In the following we describe how the SSL and TLS security protocols negotiate the cryptography algorithm that will be used before data flows between two communicating parties.

5.3.3 SSL and TLS protocols

To protect data in transit, the Secure Socket Layer (SSL) protocol [14] developed by Netscape Communications has been used extensively. SSL is a client–server security protocol that runs on top of TCP/IP. SSL provides for data encryption, and therefore addresses confidentiality and data tampering concerns as described above. One of the more common uses of SSL is secure HTTP (HTTPS) that enables the application of HTTP on top of SSL. The latest versions of Web browsers support SSL and the more recent Transport Layer Security (TLS) protocol.

The SSL protocol enables the user client and server to perform a handshake at the end of which client and server share a common secret key that is used to encrypt the information exchanged between them. A simplified depiction of the exchange of security attributes to establish a secure connection is shown in Figure 5.5 (some messages and data elements are omitted for clarity). The common secret key is the same as the session key used in conjunction with the public-key cryptography described previously. Initially, the server sends a public key to the client for encrypting client messages. The client chooses a pre-master key, encrypts it, and then sends the encrypted key to the

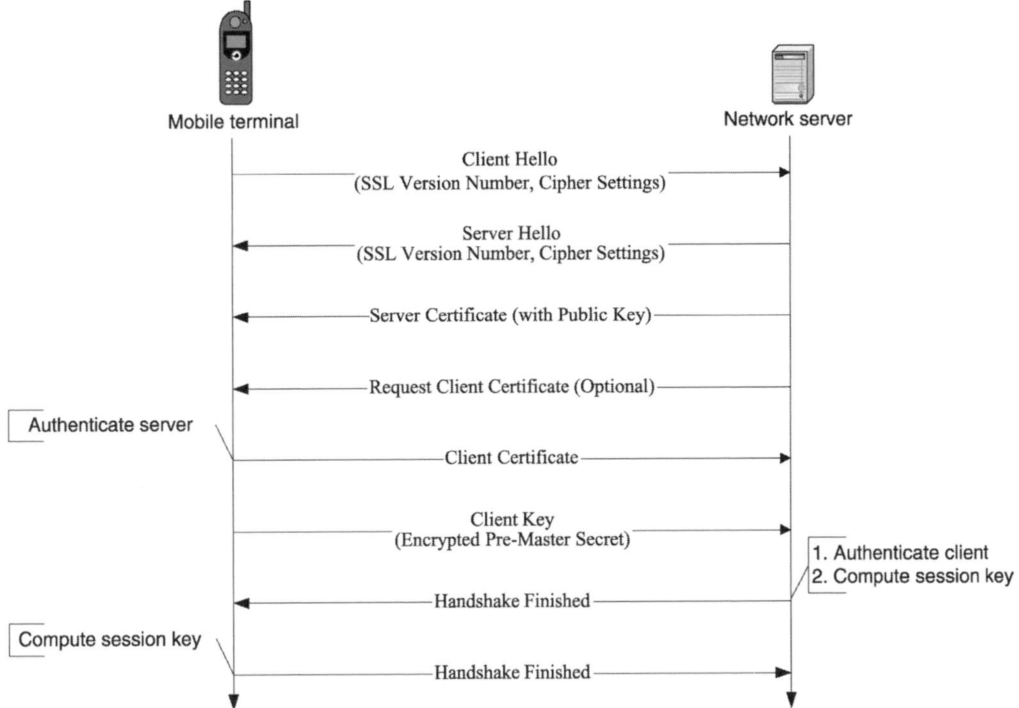

Figure 5.5 SSL handshake main messages.

server. Upon receipt of the pre-master key, both ends can compute a shared master key, using a shared algorithm, after which a session is established between client and server for confidential data exchange. During the handshake process, the server can send a server certificate (that contains its public key) to the client, and the client in turn can send a client certificate to the server. This ensures mutual authentication. Certificates are further described in a following section on certificate-based authentication.

More recently the IETF proposed the Transport Layer Security protocol (TLS), published as RFC 2246 [15] and extended with RFC 3456 [16]. TLS is based on SSL and provides essentially the same security facilities as SSL, however it is not interoperable with SSL. The original TLS RFC was extended with RFC 3546 to accommodate wireless environments that have constraints not commonly present in wired environments. These constraints include bandwidth limitations, computational power limitations, memory limitations, and battery life limitations. For example, a client mobile terminal can negotiate with a network server a maximum fragment length for exchanging application data that is smaller than the originally specified length of 16 384 bytes. Another improvement is to allow client mobile terminals to send in only a URL to their certificate, when they authenticate themselves to a server, rather than storing the certificate in the terminal and sending it upon the server's request.

Mobile terminal browsers can establish a secure SSL link between client and server if they do not require intermediate gateways for communicating with content providers. This is the case in the mobile network i-mode architecture, and browsers such as the Access NetFront microbrowser [17] support SSL. Most recent browsers support both SSL and TLS, for example Openwave's Mobile Browser 6.2 suppports SSL 3.0 and TLS 1.0.

The end-to-end secure link that is provided by SSL and TLS and is available in implementations of the WAP 2.0 architecture, as well as in the i-mode architecture, is to be contrasted with the Wireless Transport Layer Security (WTLS) that is provided in WAP 1.x between the terminal device and the wireless gateway [18]. The gateway would implement SSL or TLS to the origin content server, however, since the gateway converts between two security protocols, WTLS and SSL or TLS, data will be temporarily unencrypted during this step.

5.4 End-point access protection

In addition to protecting the data en route, the end points of a communication need to be protected from unauthorized access. A mobile terminal wants to be assured that it is connecting with the correct content server, and the content server wants to ascertain that the requesting mobile is in fact authorized to access its content. In the following we describe authentication based on passwords and a stronger scheme that uses certificates. We conclude with a description of an authorization framework specified by the IETF's AAA working group and its application in a wireless environment.

5.4.1 Password-based authentication

The IETF has addressed basic security mechanisms that can be used to protect the end-points of a communication link. The most basic authentication was defined in RFC 2617 [19]. The user's Web browser sends to a website its credentials in a GET HTTP message when it requests, for example, a document. The credentials, a "user name" and "password", are included in an authorization header. If the credentials are not valid, then a '401 unauthorized' HTTP response is returned by the service provider. Often, a user may need to track multiple passwords, one per accessed application, and a database that stores user passwords is a single point of failure for network access. To overcome the limitations of password-based solutions, "one-time sign-on" and "single sign-on" solutions have been devised.

One-time sign-on

This solution typically requires users to carry with them a physical token that generates a random password. The same password is independently generated on the server side

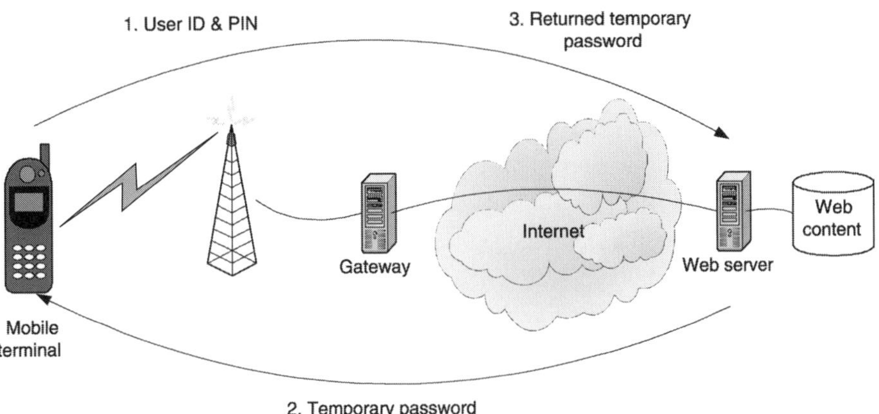

Figure 5.6 Mobile user one-time sign-on.

so that the server can perform user authentication. This is referred to as a two-factor solution as the user requires an additional device for authentication. Cell phone-based solutions that leverage this method have also been devised, where there is no requirement for a separate physical token, and the user's phone is used as the authentication device [20]. A one-time password is sent to the mobile user's phone after the user has input a user ID and personal identification number (PIN). The three-step procedure for signing on is shown in Figure 5.6. The one-time password itself expires after a few minutes, to limit the possibility of password theft.

Single Sign-On

In a distributed system, a user may invoke applications that execute on different platforms with different security mechanisms. Often the user may be required to enter a new user name and password for each platform. To alleviate this user inconvenience, a single sign-on method was devised by which users enter their credentials once, and every user access to a new domain triggers a transparent retrieval and check of the original credentials. This approach also eases the burden of system administrators and overall system security as only one set of credentials per user needs to be maintained. Also, Web services, by the nature of their distributed architecture, require a single sign-on solution that was adopted in the OASIS SAML standard described in a later section.

Typically, when a user wants to access an application, the user's client software will access a central server that stores user credentials. The central server delivers the user credentials to the user's client, and these credentials are then sent to the application for user verification. Different architectures can be used to support single sign-on. When users access a secondary domain, their credentials are checked:

- Directly, if the primary domain passes on the user credentials to the secondary domain at the time of secondary domain access. For example, the user client terminal will forward the locally cached user credentials to a secondary domain's application upon access to this latter application.

- Indirectly, if user credentials are retrieved from an information management system of single sign-on credentials. For example, the secondary domain's application issues a request for user credentials to the central credentials server.
- Immediately, if the secondary domain applications received the user credentials at the time of primary domain sign-on.

Specifications and supporting products for single sign-on were produced by Liberty Alliance member companies that include, among others, Sun, Nokia, Ericsson, and NTT DoCoMo [21]. The Liberty Alliance was established in 2001 by companies with a common goal of creating standards for federated identity management. Federated identity allows users to link identity information between accounts without centrally storing personal information. In practice, this means that users can be authenticated by one company or website, and don't need to sign on with a separate username and password at other Liberty-enabled sites that are part of a common circle of trust. The first phase of the Liberty specifications was released in 2002 and the second phase, which defines a framework for creating, discovering, and consuming identity services, was released in 2003.

5.4.2 Certificate-based authentication

When compared to password-based authentication, a stronger user identification method is certificate-based authentication introduced by Kohnfelder [22]. Certificate authorities, similar to how social security cards are issued by the US Social Security Administration office, issue "digital certificates", also referred to as digital IDs. The ITU-T standards body defined certificate contents in recommendation X.509 [23]. A certificate's information consists of the user's public key, the cryptographic algorithm used, the user's name, an expiration date, the name of the issuing certificate authority, the certificate authority's signature (an identifier of the encryption and hash function used to sign the certificate), etc. Certificates that contain a user's public key are referred to as "public-key certificates".

Certificates cannot be forged since they are digitally signed by issuing certificate authorities, for example, VeriSign [24], and a receiver uses a certificate authority's well-known public key to decrypt a certificate. In a client and server model, the mobile client software sends the network server a certificate, or a URL pointing to a network-stored certificate, in addition to a digitally signed message (Figure 5.7). Upon receipt of the certificate, the server can extract from it the client's public key for decrypting the client's message. The binding between the client's name and client's public key is used to authenticate the client. This authentication scheme is more automated compared to the password-based approach, as no direct user involvement is needed.

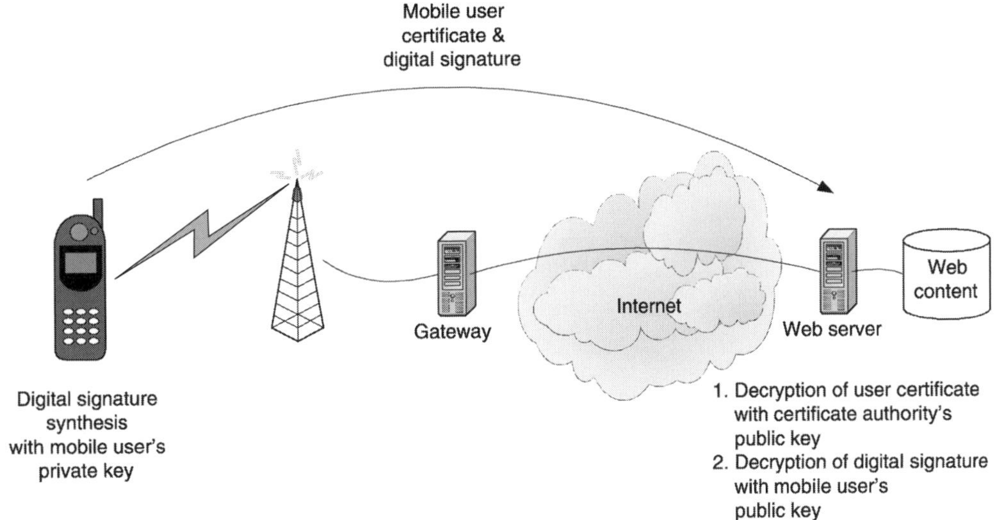

Figure 5.7 Mobile user authentication with user's certificate.

How can the server be assured that the certificate was indeed sent by the client in question? To reinforce confidence that the certificate's public key belongs in fact to the sender, or in other words, that the public key identifies the sender, sometimes more than one certificate is included in the message. These certificates form a "certificate chain", where the highest-level certificate is issued by an authority well known to the receiver, and is therefore a trusted authority. In a certificate chain each certificate attests that the binding between the identity of the sender and the associated public key is valid since a higher-level authority signs the certificate. Certificate verification starts from the bottom of the chain and proceeds up to the highest-level certificate. Many checks are required to validate a certificate chain. However, if two parties communicate often with each other then the validation needs to be performed once and after that the receiver stores the public key of the sender and uses it to decrypt future sent messages. Another concern with certificates is that the private key of a certificate sender needs to be stored securely to prevent others from impersonating the sender. For example, a hard tamper-resistant apparatus on the mobile terminal can be used to store the private key.

A mobile terminal can use the same scheme to authenticate a network server and be assured that it is communicating with the right entity. This authentication check is applicable in both mobile initiated and server initiated communications. For example, in a server initiated information push operation, the mobile terminal may want to authenticate the server prior to agreeing to establish the connection to thwart unwanted wireless advertisements.

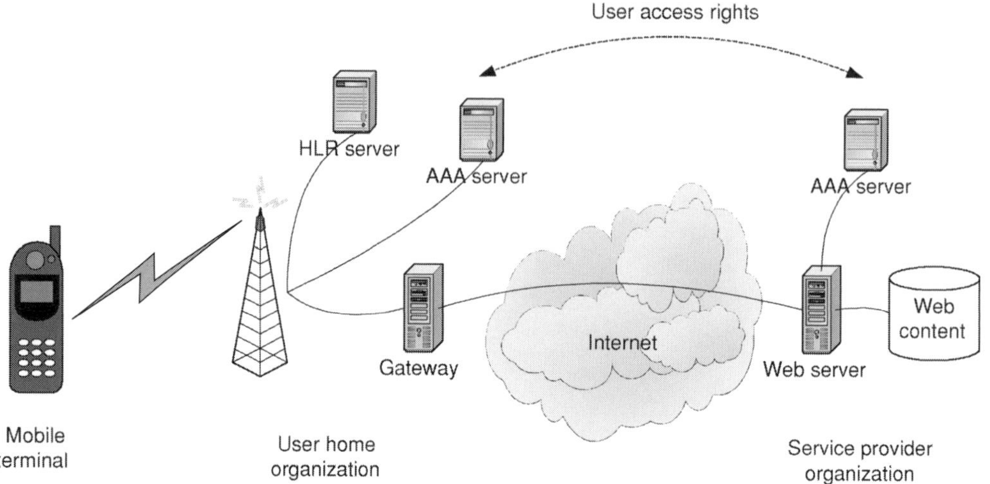

Figure 5.8 Basic authorization network elements.

5.4.3 Authorization framework

Authorization is the process by which an authority grants a requestor access to resources. Resources could be, for example, data stores or services that can be invoked by a user. The IETF's authentication, authorization, and accounting (AAA) working group (WG) has specified the components of an authorization architecture [25] and [26]. An authorization process, as defined by the AAA WG, typically involves a user (could be a program), a users' home organization with its AAA server, and a service providers' organization that contains both the providers' AAA server and the providers' resources. Figure 5.8 shows the associated network elements when the user's home organization is the mobile user's cellular operator and the service provider is an Internet content provider. The mobile user authentication and authorization information for the cellular network is stored in the network's home location register (HLR). In the following, we use the AAA WG terminology, and, to prevent confusion, we substitute in some cases terms that are commonly applied in the wireless industry.

Authorization is usually based on agreements between the participating entities. The user will have an agreement set with the home AAA server, for example, an agreement on the rates of service access as a function of the time of day. Separate agreements will be set between the home AAA server and the service provider's AAA server, for example, the bandwidth of data transfer between the home organization and the service provider organization, or agreements that determine access constraints.

In a typical set-up, the user home organization will be separate from the service provider organization. A mobile user's resource access rights in a service provider's organization could be established by message exchange between the home and service provider AAA servers (Figure 5.8). To complicate this further, the service provider

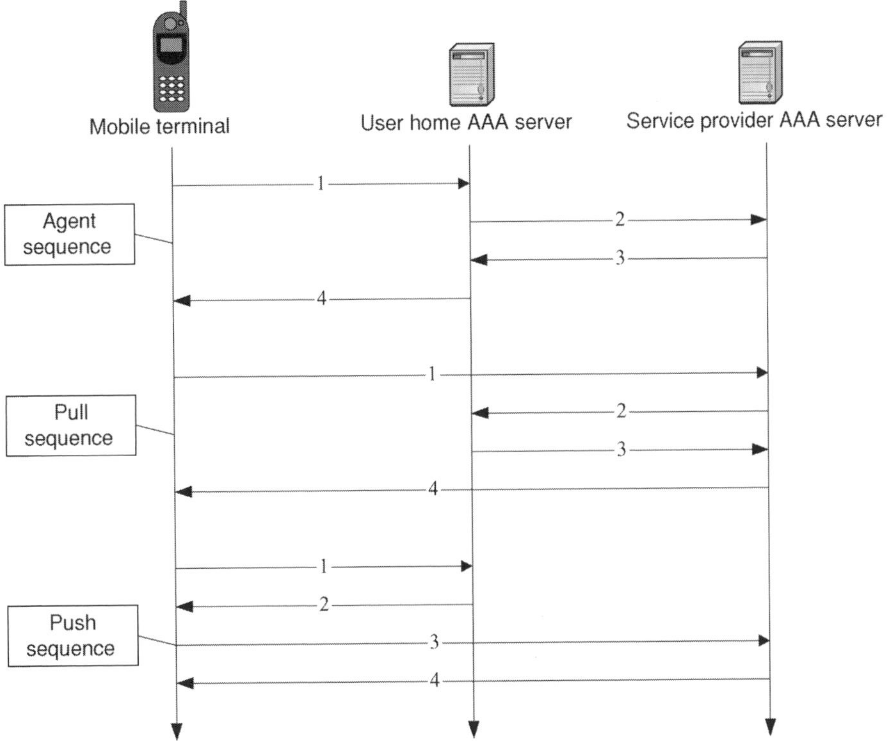

Figure 5.9 Message sequences for authorization verification.

may request the support of another service provider to provide a "distributed service". This will require that the AAA servers of the service providers exchange corresponding messages to establish the resource access authorization rights. An alternative approach to the described scheme of peer-to-peer communication between the AAA servers is for a central AAA broker to enable the communication between disparate servers. In this case, each AAA server will need to support just one communication protocol with the central AAA broker.

Authorization assessment procedures

The IETF has defined three main message sequences for an authorization assessment procedure to verify a user's access rights [26]. These are termed, "agent", "pull", and "push", and are illustrated in Figure 5.9. In the agent case, the user requests services from the service provider through the AAA server of the user's home domain. The home AAA server forwards the user request to the service provider who can be in another domain. In the pull case, the service provider, upon receiving a request for service from the user, will query the user's home AAA server. Finally, in the push case, the user first gets from the home AAA server an assertion that validates the user's capability to access the service. Then, the user includes this assertion in the service request sent to the service provider.

The following example shows how to apply authorization schemes in the case of a mobile user that visits a museum and wishes to access museum content information through a local wireless network installed in the museum premises. As the museum's content server is typically owned and operated by a domain different from the user's home organization, authorization to access the content server could proceed in the following ways:

- The mobile user access request is forwarded to the museum's AAA server, and the latter then accesses the user's home AAA server to verify authorization to access the museum's content server. This is the pull authorization procedure described above.
- The mobile user first retrieves an authorization token from its home AAA server. This token is then included in the request for information sent to the museum's AAA server to authorize access to the museum's content server. This is the push authorization procedure described above.

After the mobile user is granted access, the museum's content server may want to access the user's profile available in the user's home domain in order to tailor the information delivered to the wireless terminal. In this case, the content server is the entity that issues the access request, and its authorization needs to be verified. Authorization to access the user's profile could proceed in the following ways:

- The content server, acting on behalf of the mobile user, accesses the museum's AAA server, and the latter then accesses the user's home AAA server to verify authorization to access the user profile. This is the agent authorization procedure described above.
- The content server first retrieves an authorization token from the museum's AAA server. This token is then included in the request sent to the user's home AAA server. The latter server uses this token to verify access to the user's profile. This is the push authorization procedure described above.

Other examples of authorization application are described in [27]. An often occurring case where mobile user access needs to be authorized, is when mobile users roam across wireless networks that are in different administrative domains. Mobility management is addressed by the IETF Mobile IP working group [17], and the AAA working group [25] recommended authorization solution follows.

When a mobile terminal attempts to access a foreign network, it forwards its Network Access Identifier (NAI) [28], to the foreign AAA server. The NAI is of the form "user@realm", and the foreign AAA server, or a broker, uses the "realm" to identify the mobile terminal's home AAA server. The authorization request is forwarded from the foreign AAA server to the home AAA server. This is the pull authorization procedure. Upon successful processing of the access authorization, communication links are established between the following three entities: the mobile terminal, a home agent to receive messages destined to the mobile terminal, and a foreign agent that connects the mobile terminal to the foreign network.

Attribute certificates

In the above authorization models, the assertion for service access authorization is retrieved from an AAA server. An alternative to these models is to have the requestor use a certificate that lists access control information that can include elements such as time validity of the certificate and a client's role assignment, as defined in ITU-T X.509 [23]. This certificate is referred to as an "attribute certificate". Use of an attribute certificate presupposes that the user was authenticated with a public-key certificate that is referred to in the attribute certificate.

An attribute certificate lists the privileges that an entity has for accessing a resource, and is digitally signed by an attribute authority. It can be stored in a publicly accessible repository, and retrieved later by servers that need to make an authorization decision. Alternatively, an entity may request an attribute certificate from an attribute authority, and later supply it to any server that stores a protected resource.

5.5 Web services security

For ease of component integration reasons, the wireless network infrastructure that supports wireless Web information services would typically be implemented with Web services technology. Web services handle user requests for information and provide corresponding responses, and as such, access to the Web services, as well as their data transmissions, needs to be secured. The following sections focus on how to secure Web services.

Web services messages are composed of XML to ensure portability across platforms and operating systems. XML is ASCII text and as such can be easily intercepted by any server on the way between the sender and receiver. XML messages typically require processing at intermediary sites, and the underlying protocols, namely SOAP and HTTP, do not include security features. In addition, Web sites expose the Web services methods that can be called on their site.

An example scenario of XML message processing is when a mobile user submits a purchase order from a mobile terminal with an *orderID* included in the order's header. This message is processed by an order processing subsystem that may append a *shippingID* to the header. This order is then shipped to the shipping department that appends a *shippedInfo* to the header. The message is then sent to the billing department for processing. Three separate organizations, the mobile user, the order processing subsystem, and the shipping department, processed and modified an XML message in this example.

The mobile user signs the *orderID* header and the body of the request (the contents of the order). The order subsystem would then sign, at a minimum, the *orderID* and the *shippingID*, and possibly the body as well. After this order is processed and shipped by the shipping department, the shipping department would sign, at a minimum, the *shippedInfo* and the *shippingID*, and possibly the body, and forward the message to the billing department. The billing department can verify the signatures and determine a valid chain of trust for the purchase order, as well as who did what.

The above scenario illustrates why, due to the multiplicity of involved sites, transport-level security by itself is insufficient. Initial security implementations have used proprietary solutions; however, there are a number of standards that cover the two main areas addressed above, namely data transmission and end-point access protection.

Data transmission protection is handled by:

- **XML Encryption** [8]. A W3C standard that specifies how a complete or parts of an XML document can be encrypted.
- **XML Signature** [29]. A W3C standard that specifies how a complete or parts of an XML document can be digitally signed. By validating the signature, the receiver is assured that the data was not altered en route.
- **XKMS (XML Key Management Specification)** [30]. A W3C standard that specifies how Public Key Infrastructure (PKI) is used to provide XML-based clients with cryptographic keys.
- **WS-Security** [31], [32], [33]. An industry specification that provides for secure SOAP message exchange.

End-point access protection is handled by:

- **SAML (Security Assertion Markup Language)** [34]. An OASIS specification that provides for a single sign-on to authenticate users and authorize access to multiple sites.
- **XML Signature** [29]. A W3C standard that also addresses user authentication. A valid signature assures the receiver that the requestor is in fact who it claims to be.

5.5.1 XML security

The W3C, working jointly with the IETF, has produced specifications that address the security of XML documents.

XML encryption

XML Encryption is a recommendation that enables the encryption of a complete XML document, of selective XML elements, or of XML elements' content [8]. For example, a mobile user could encrypt a purchase order to ensure that the data is protected while in transit to a website. The representation of an encrypted XML element is shown in Listing 5.1.

```
[1]  <EncryptedData Type='http://www.w3.org/2001/04/xmlenc#Element'
[2]    xmlns='http://www.w3.org/2001/04/xmlenc#'>
[3]   <CipherData>
[4]     <CipherValue>A23B45C56</CipherValue>
[5]   </CipherData>
[6]  </EncryptedData>
```

Listing 5.1 Encrypted XML element

The *<EncryptedData>* element is the representation of an encrypted element within an XML document that may have other non-encrypted components. The *<CipherData>* element can either envelop or reference the raw encrypted data that results from the application of a cipher. In the first case, shown in Listing 5.1, that encrypted data is within the *<CipherValue>* element. Alternatively, a *<CipherReference>* element contains a URI attribute that points to the location of the encrypted data.

XML signature

The XML Signature recommendation addresses the signing of XML content [29]. In addition to the encryption step described above, the mobile user could sign the purchase order to ensure that it is not tampered with, and to enable the website receiver to authenticate the user. Listing 5.2 provides an abbreviated example signature from the XML Signature recommendation.

```
<Signature Id=“MyFirstSignature”
  xmlns=“http://www.w3.org/2000/09/xmldsig#”>
  <SignedInfo>
   <SignatureMethod
    Algorithm=“http://www.w3.org/2000/09/xmldsig#dsa-sha1”/>
   <Reference
    URI=“http://www.w3.org/TR/2000/REC-xhtml1-20000126/”>
    <DigestMethod
     Algorithm=“http://www.w3.org/2000/09/xmldsig#sha1”/>
    <DigestValue>j6lwx3rvEPO0vKtMup4NbeVu8nk=</DigestValue>
   </Reference>
  </SignedInfo>
  <SignatureValue>MC0CFFrVLtRlk=...</SignatureValue>
  <KeyInfo>
   <KeyValue>
    <DSAKeyValue>
     <P>...</P><Q>...</Q><G>...</G><Y>...</Y>
    </DSAKeyValue>
   </KeyValue>
  </KeyInfo>
</Signature>
```

Listing 5.2 Signed XML content

The *<SignedInfo>* element contains the source data that is actually signed or a reference to that data. In the present case a reference is provided, and the resulting signature is referred to as a detached signature. The *<SignatureMethod>* element includes the URI of the algorithm used to derive the signature; this algorithm is a combination of the digest algorithm and a key-based encryption algorithm. The *<Reference>* element has a *URI* attribute that points to the location of the source data that is signed. The enclosed *<DigestMethod>* contains the URI of the digest algorithm, for example a hashing algorithm, which generates a concise representation of the source data that is provided in *<DigestValue>*. The next element, *<KeyInfo>*, is optional and identifies the public key used for encrypting and validating the signature. This key could be in the form of a certificate, and is optional since key information could be already available to the receiver. Finally, the signature itself is found in the *<SignatureValue>* element.

XML key management

The XML Key Management specification (XKMS) defines services and messages for querying about and registering security keys: these are, respectively, the key information service specification (X-KISS) and the key registration service specification (X-KRSS) [30]. These services were defined in part to support client terminals with constrained processing capabilities, such as cell phone devices, as they enable the offloading of sophisticated key management functionality.

By design, the XML Signature specification does not mandate use of a particular trust policy. The signer of a document is not required to include any key information but may include a *<KeyInfo>* element that specifies, for example, the key itself, a key name, or an ITU-T X.509 certificate [23]. The information provided by the signer may therefore be insufficient by itself to perform cryptographic verification and decide whether to trust the signing key, or the information may not be in a format the client can use. For example, the key may be specified by name only, or the key may be encoded in an X.509 certificate that the client cannot parse.

The X-KISS key information service allows a client to delegate part or all of the tasks required to process XML Signature key information elements (*<KeyInfo>* in Listing 5.2) to an XKMS service. A key objective of the XKMS service is to minimize the complexity of clients using XML Signature. An XKMS client is relieved of the complexity and syntax of the underlying public key infrastructure used to establish trust relationships, which may be based upon a specification such as ITU-T's X.509. For encryption operations, a client that does not know the public key of a recipient may also query an XKMS service for this key.

Listing 5.3, from the XKMS specification [30], shows an example request issued to an XKMS service when the client terminal receives a signed document with a corresponding certificate, but does not have the capability to extract the public key from the certificate. The client's request is contained in the *<LocateRequest>* element which encloses the sender's certificate in the *<KeyInfo>* element. The request specifies in the *<RespondWith>* element that the service returns the sender's public key value.

```
<? xml version="1.0" encoding="utf-8"?>
<LocateRequest xmlns:ds=
  "http://www.w3.org/2000/09/xmldsig#"
  xmlns:xenc="http://www.w3.org/2001/04/xmlenc#"
```

```
[4]    Id="I4593b8d4b6bd9ae7262560b5de1016bc"
[5]    Service="http://test.xmltrustcenter.org/XKMS"
[6]    xmlns="http://www.w3.org/2002/03/xkms#">
[7]   <RespondWith>KeyValue</RespondWith>
[8]   <QueryKeyBinding>
[9]    <ds:KeyInfo>
[10]     <ds:X509Data>
[11]<ds:X509Certificate>MIICAjCCAW+gAwIBAgIQlzQovIEbLLh
      Ma8K5MR/juzAJBgUrDgMCHQUAMBIxEDAOBgNVBAMTB1Rlc3Qg
      Q0EwHhcNMDIwNjEzMjEzMzQxWhcNMzkxMjMxMjM1O
      TU5WjAsMSowKAYDVQQGEyFVUyBPPUFsaWNlIENvcnAgQ049
      QWxpY2UgQWFyZHZhcmswgZ8wDQYJKoZIhvcNAQEBBQADgY0A
      MIGJAoGBAMoy4c9+NoNJvJUnV8pqPByGb4FOJcU0VktbGJpO2im
      iQx+EJsCt27z/pVUDrexTyctCWbeqR5a40JCQmvNmRUfg2d8
      1HXyA+iYPl4L6nUlHbkLjrhPPtMDSd5YHjyvnCN454+Hr0
      paA1MJXKuw8ZMkjGYsr4fSYpPELOH5PDJEBAgMBAAGjRzBFMEMGA1
      UdAQQ8MDqAEEVr1g8cxzEkdMX4GAlD6TahFDASMRAwDgY
      DVQQDEwdUZXN0IENBghBysVHEiNFiiE2lxWvmJYeSM
      AkGBSsOAwIdBQADgYEAKp+RKhDMIVIbooSNco
      IeV/wVew1bPVkEDOUwmhAdRXUA94uRifiF
      fmp9GoN08Jkurx/gF18RFB/7oLrV
      Y+cpzRoCipcnAnmh0hGY8FNFmhyKU1tFhVFdFXB5QUglkmkRntN
      kOmcb8O87xO0XktmvNzcJDes9PMNxrVtChzjaFAE
      =</ds:X509Certificate>
[12]     </ds:X509Data>
[13]    </ds:KeyInfo>
[14]    <KeyUsage>Signature</KeyUsage>
[15]   </QueryKeyBinding>
[16]  </LocateRequest>
```

Listing 5.3 XKMS request for certificate processing

The XKMS service response to the client terminal includes the public key in the *<RSAKeyValue>* element in Listing 5.4.

```
[1]   <?xml version="1.0" encoding="utf-8"?>
[2]   <LocateResult xmlns:ds=
        "http://www.w3.org/2000/09/xmldsig#"
[3]      xmlns:xenc="http://www.w3.org/2001/04/xmlenc#"
[4]      Id="I46ee58f131435361d1e51545de10a9aa"
[5]      Service="http://test.xmltrustcenter.org/XKMS"
         ResultMajor="Success"
[6]      RequestId="#I4593b8d4b6bd9ae7262560b5de1016bc"
```

```
      xmlns="http://www.w3.org/2002/03/xkms#">
  <UnverifiedKeyBinding Id=
    "I36b45b969a9020dbe1da2cb793016117">
    <ds:KeyInfo>
      <ds:KeyValue>
        <ds:RSAKeyValue>
<ds:Modulus>zvbTdKsTprGAKJdgi7ulDR0eQB
    ptLv/SJNIh3uVmPBObZFsLbqPwo5nyLOkzWlEHNbShPMRp1qFrAfF13
    LMmeohNY fCXTHLqH1MaMOm+BhXABHB9rUKaGoOBjQPHCBtHbfM
    GQYjznGTpfCdTrUgq8VNl qM2Ph9XWMcc7qbjNHw8=</ds:Modulus>
          <ds:Exponent>AQAB</ds:Exponent>
        </ds:RSAKeyValue>
      </ds:KeyValue>
    </ds:KeyInfo>
    <KeyUsage>Signature</KeyUsage>
    <KeyUsage>Encryption</KeyUsage>
    <KeyUsage>Exchange</KeyUsage>
    <UseKeyWith Application="urn:ietf:rfc:2633"
      Identifier="alice@alicecorp.test"/>
  </UnverifiedKeyBinding>
</LocateResult>
```

Listing 5.4 XKMS response with public key

The X-KRSS key registration service describes a protocol for registration and subsequent management of public key information. A client may request that the registration service bind information to a public-key pair, that is, the private key and the public key. The information bound may include a name, an identifier, or extended attributes defined by the implementation. The key pair to which the information is bound may be generated in advance by the client or on request generated by the service. If the private key is used for signing, it is usually preferable to have the client generate the public key pair. The registration protocol may also be used for subsequent management operations including recovery of the private key and reissue or revocation of the key binding.

5.5.2 WS-security

In April 2002, IBM, Microsoft, and Verisign, published WS-Security to help enterprises build secure and interoperable Web services. These companies have also outlined a roadmap for additional Web services security specifications ([31], [32], [33]) that address policy, trust, privacy, secure conversation, federation (for managing trust relationships), and authorization. The OASIS industry forum has since taken up WS-Security in the form of the SAML specification described in section 5.5.3.

WS-Security defines a standard set of SOAP extensions in the form of headers that implement message integrity, confidentiality, and authentication. These headers are in fact security tokens associated with messages that could be implemented in various forms such as ITU-T X.509 certificates, IETF Kerberos tickets [35], or username and password combinations. WS-Security is a foundation that can serve as a basis for building a wide variety of security models that include PKI, Kerberos, and SSL. Therefore, this specification does not specify explicit security protocols but defines security tokens that when combined with digital signatures can serve to authenticate senders. An example security token is a signed certificate that binds a sender's identity with a public key. In addition, WS-Security provides for message confidentiality through encryption by using XML Encryption, and message integrity by using XML Signature in conjunction with security tokens.

5.5.3 SAML

The Security Assertion Markup Language (SAML) is an OASIS standard produced by the Security Services Technical Committee, which specifies an XML-based framework for Web services that allows authentication, authorization, and entity-attribute information to be exchanged among different Web sites [34]. SAML depends on the availability of authentication frameworks such as PKI digital certificates. It includes Single Sign-On (SSO) by a user or computer, as defined in the SAML profiles specifications, so that security assertions can hold both within and across security domains. An example benefit of SSO is that visitors can use sites hosted by multiple companies, making it easier for online shoppers without requiring them to log in and re-authenticate individually at each site.

SAML incorporates industry-standard protocols and messaging frameworks, such as XML Signature, XML Encryption, and SOAP. The SAML security information is expressed as assertions about entities such as persons and computers, and these assertions convey information about entity authentication, authorization to access resources, and entity attribute values. SAML assertions are issued by SAML authorities such as "authentication authorities" and "attribute authorities". For example, the *<AuthenticationStatement>* element issued by a SAML authentication authority states that its subject was authenticated by a particular means at a particular time. The fragment schema in Listing 5.5 from the SAML standard defines this element.

```
[1]  <element name="AuthenticationStatement"
[2]   type="saml:AuthenticationStatementType"/>
[3]  <complexType name="AuthenticationStatementType">
[4]   <complexContent>
```

```
      <extension base="saml:SubjectStatementAbstractType">
        <sequence>
          <element ref="saml:SubjectLocality" minOccurs="0"/>
          <element ref="saml:AuthorityBinding"
            minOccurs="0" maxOccurs="unbounded"/>
        </sequence>
        <attribute name="AuthenticationMethod" type="anyURI"
          use="required"/>
        <attribute name="AuthenticationInstant" type="dateTime"
          use="required"/>
      </extension>
    </complexContent>
  </complexType>
```

Listing 5.5 SAML Authentication Statement schema

The optional *<SubjectLocality>* element defines the domain name server (DNS) and the IP address of the authenticated subject, and the optional *<AuthorityBinding>* element specifies additional SAML authorities that may be available to provide additional information about the authenticated subject. The *<AuthenticationMethod>* element provides a URI reference to the type of authentication, for example, password, Kerberos, SSL or TLS certificate-based authentication, X.509 public key, XML signature, etc. Finally, *<AuthenticationInstant>* specifies the time at which authentication took place.

SAML also defines an XML-based request and response protocol to request assertions from SAML authorities. The protocol can be bound to underlying transports such as SOAP over HTTP. The previously described XKMS request and response protocol for security key management was designed to be compatible with the SAML standard. The requestor specifies what type of assertions is desired from the SAML authority for a specific subject. For example, the request can include XML elements such as:

- *<AuthenticationQuery>* to query for authentication information about authentication acts that have occurred in a previous interaction between the indicated subject and the Authentication Authority.
- *<AuthorizationDecisionQuery>* to request an authorization decision on one or more specified actions on a resource.
- *<AttributeQuery>* to query for attribute information on the indicated subject.

The authorization element, in an authorization query, will include the following attribute and element:

- *Resource*: a URI that indicates the resource for which authorization is requested.
- *<Action>*: one or more requested actions on the resource such as read, write, execute, delete, and control.

Optional elements can be included in a request; for example, to filter possible responses the authentication element *<AuthenticationQuery>* may contain an XML element *<AuthenticationMethod>* that, as in the above authentication statement, specifies how authentication was done. In this case, at least one element in the set of returned authentication assertions must match the specified authentication method.

Any SAML request, response, or even assertion, may be signed to ensure message integrity and authentication of the message originator. The signature also provides for non-repudiation of origin if the signature is based on the originator's public–private key pair. The signature is not a SAML requirement when another technique is used for direct authentication and establishment of a secure channel between communicating parties, for example, SSL, or password-based login.

5.6 Industry public-key cryptography standards

Among the security industry de facto standards, the Public-Key Cryptography Standards (PKCS) specifications produced by RSA Laboratories have become widely referenced and implemented in many standards (for example, SSL) [36]. The PKCS documents were generated for the purpose of accelerating the deployment of public-key cryptography, and were first published in 1991.

Many mobile products have implemented RSA's PKCS specifications. For example, the Access NetFront v3.1 micro-browser includes support for PKCS #7, #10, #11, and #12, which enable the inclusion of mobile terminals in security models [37]. PKCS #7 provides syntax for data that may have cryptography applied to it, and PKCS #10 describes a standard for certification requests sent to a Certificate Authority to issue certificates directly to the mobile terminal. The public and private keys can be generated at the mobile terminal for maximum security. This ensures the private key never leaves the client terminal. PKCS #11 specifies an API to a hardware token such as a smart card, and PKCS #12 describes a syntax for transfer of personal identity information, including private keys, certificates, and miscellaneous secrets.

REFERENCES AND FURTHER READING

[1] 3GPP, *Security Architecture*, TS 33.102. (Sep. 2003).

[2] 3GPP, *Generic Authentication Architecture*, TS 33.919. (Dec. 2003).

[3] 3GPP, *Generic Bootstrapping Architecture*, TS 33.220. (Dec. 2003).

[4] T. Hiller *et al., CDMA2000 wireless data requirements for AAA*, RFC 3141. IETF (Jun. 2001).

[5] T. Natsuno, *i-mode Strategy* (Chichester, England: John Wiley and Sons Ltd, 2003).

[6] RSA Security, http://www.rsasecurity.com.

[7] NIST, *Data Encryption Standard (DES)*, Federal Information Processing Standard Publication 46 (FIPS PUB 46), 1977.

[8] T. Imamura *et al., XML Encryption Syntax and Processing*. W3C Recommendation (Dec. 2002), http://www.w3.org/TR/xmlenc-core/.

[9] W. Diffie and M. Hellman, New directions in cryptography. *IEEE Trans. Information Theory*, **22**:6 (1976), 644–54.

[10] R. L. Rivest *et al.*, A method for obtaining digital signatures and public-key cryptosystems. *Comm. of the ACM*, **21**:2 (1978), 120–6.

[11] Netscape Communications, *Introduction to SSL*. (http://developer.netscape.com/docs/manuals/security/sslin/contents.htm, 1998).

[12] R. Rivest, *The MD5 Message-Digest algorithm*, RFC 1321. IETF (Apr. 1992).

[13] NIST, *Secure Hash Standard*, FIPS PUB 180–1, Apr. 1995.

[14] A. O. Freier *et al., The SSL Protocol v3.0.* (Netscape Communications, Mar. 1996).

[15] T. Dierke and C. Allen, *The TLS protocol vl.0*, RFC 2246. IETF (Jan. 1999).

[16] S. Blake-Wilson *et al., TLS extensions*, RFC 3456. IETF (Jun. 2003).

[17] Access Co., http://www.access.co.jp/english/company/index.html.

[18] WAP Forum, *Wireless Transport Layer Security*. WAP-261-WTLS-20010406-a. (http://www.openmobilealliance.org/tech/affiliates/wap/wapindex.html, Apr. 6, 2001).

[19] J. Franks *et al., HTTP authentication: Basic and digest access authentication*, RFC 2617. IETF (Jun. 1999).

[20] RSA Security, *RSA Mobile*, http://www.rsasecurity.com/node.asp?id=1314.

[21] Liberty Alliance Project, http://projectliberty.org/.

[22] L. M. Kohnfelder, Towards a practical public-key cryptosystem. S. B. Thesis, MIT (May 1978).

[23] ITU-T, *The Directory: Public-key and Attribute Certificate Frameworks*, X.509, Mar. 2000.

[24] Verisign, http://www.verisign.com.

[25] C. de Laat *et al., Generic AAA architecture*, RFC 2903. IETF (Aug. 2000).

[26] J. Vollbrecht *et al., AAA authorization framework*, RFC 29004. IETF (Aug. 2000).

[27] J. Vollbrecht *et al., AAA authorization application examples*, RFC 2905. IETF (Aug. 2000).

[28] B. Aboba and M. Beadles, *The network access identifier*, RFC 2486. IETF (Jan. 1999).

[29] M. Bartel *et al., XML-Signature Syntax and Processing*. W3C Recommendation (Feb. 2002), http://www.w3.org/TR/xmldsig-core/.

[30] P. Hallam-Baker *et al., XML Key Management Specification (XKMS) v2.0.* W3C Working Draft (Apr. 2003), http://www.w3.org/TR/xkms2/.

[31] IBM, *Specification: Web Services Security (WS-Security) v1.0.* (http://www.106.ibm.com/developerworks/library/ws-secure, Apr. 5, 2002).

[32] Microsoft, *WS-Security Roadmap*, http://msdn.microsoft.com/library/enus/dnglobspec/html/ws-security.asp.

[33] Verisign, *WS-Security Roadmap*, http://verisign.com/wss/wss.pdf.

[34] E. Maler *et al., Assertions and protocol for the OASIS Security Assertion Markup Language (SAML) v1.1.* OASIS (Sept. 2003), http://www.oasis-open.org/specs/index.php#samlv1.1.

[35] J. Kohl and C. Neuman, *The Kerberos network authentication service v5*, RFC 1510. IETF (Sep. 1993).

[36] RSA Security, *PKCS*, http://www.rsasecurity.com/rsalabs/node.asp?id=2124.

[37] D. Zucker *et al., Ubiquitous Web Browsing: Evolving Toward Universal Access Everywhere*, Access Co., White paper. http://www.access.co.jp, Sep. 2003.

R. Atkinson, *Security architecture for the Internet Protocol*, RFC 1825. IETF (Aug. 1995).

BNX Systems, *Enterprise Single Sign-on: Balancing Security and Productivity*, http://www.bionetrix.com.

P. Gutman, PKI: It's not dead, just resting. *IEEE Computer*, (Aug. 2002), 41–9.

C. Perkins, *IP mobility support*, RFC 2002. IETF (Oct. 1996).

Netscape Communications, *Introduction to public-key cryptography*. (http://developer.netscape.com/docs/manuals/security/pkin/index.html?content=contents.htm#1053011, 1998.)

OASIS, *Security and privacy considerations for the OASIS Security Assertion Markup Language (SAML)*. (http://www.oasis-open.org/specs/index.php#samlv1.1, May, 2002).

WAP Forum, *Public Key Infrastructure Definition*. WAP-217-WPKI. (http://www.openmobilealliance.org/tech/affiliates/wap/wapindex.html, Apr. 24, 2001).

6 Personalization and privacy

This chapter introduces the benefits of personalization of mobile interactions and the associated privacy concerns. We start by describing the objectives of personalization and what application domains can be impacted by it. Next we review how to build user behavior models as these models are a key element for designing a personalization system. We proceed by elaborating on recommender systems that tailor the Web-based information delivered to mobile users. Three major types of recommenders are described: content, collaborative, and hybrids.

We then describe the architectural components needed to realize a personalization system. These components include user profiles, where preferences are stored, and personalization rules that match between user attributes and content. Ontology mediators can play a central role to correlate between user-specified attributes and content descriptions. Finally, we conclude by addressing related privacy concerns and detail the W3C's P3P effort at answering these by catering to the privacy of browser-based clients. The P3P architecture is explained and associated policy files are presented.

6.1 Objectives of personalization

Personalization was defined by the Personalization Consortium [1], active in 2000, as the use of technology and customer information to tailor electronic commerce interactions between a business and each individual customer. This tailoring of interactions has the following stated goals:

- Better serve the customer by anticipating needs.
- Make the interaction efficient and satisfying for both parties.
- Build a relationship that encourages the customer to return for subsequent purchases.

The above goals reflect the business objectives of personalization. By knowing the customer and its needs, the business can target products and services to fit. And, the business hopes to develop customer loyalty through this more efficient customer service. Figure 6.1 shows some of the areas that can be significantly impacted by personalized interactions, and we elaborate on these in the following.

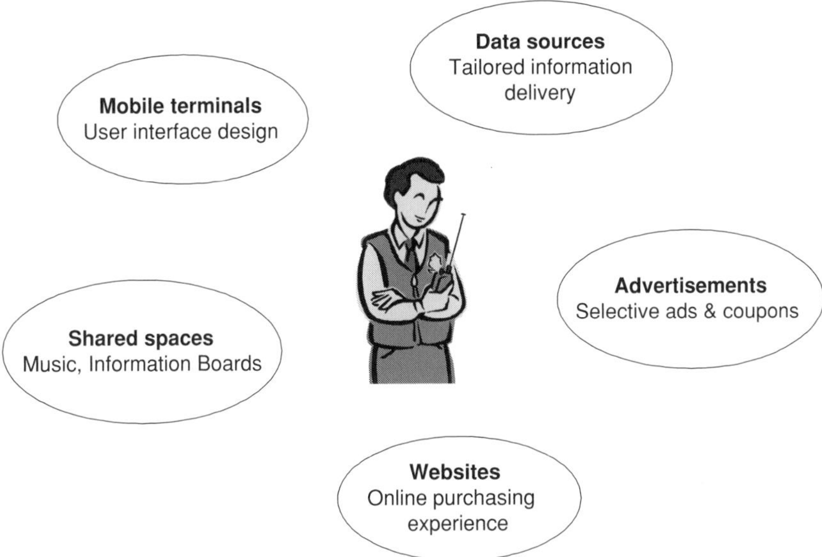

Figure 6.1 Personalization impact areas.

Companies and organizations have deemed personalization to be an important factor in developing their websites to enhance the site visit experience. For example, My Yahoo! [2], [3] allows a user to customize a Web page layout with user picked content such as news items, stocks to track, weather reports, TV listings, and traffic reports. Staples [4], an office supplies vendor, allows users to create lists of favorite items that they often purchase and be sent email reminders to replenish supplies. Another often cited example is Amazon's [5] recommendation of additional book titles to a user by comparing the user's current purchase to those of other customers with similar interests. The site visitor is presented with a message "people who bought that item also bought" similar to what a real personal assistant might suggest. This personal assistant can be made more intelligent by retrieving the user's history of book purchases and deducing the user's purchase preferences so that its recommendations are better tailored to the current user. Amazon has a total of 23 independent applications that operate at different sophistication levels to personalize visits to their website [6].

Personalization is also a critical element in the design of wireless Web information services and associated user interfaces on mobile terminals. With the added constraints of limited user time and constrained terminal input capabilities, personalization can help ease the interaction experience so that users are presented with the information they need and in the format they prefer at the outset of a wireless session. The presented information can be made compatible with the user's context. For example, a user starting a car ride can be presented with navigation data that takes into account the end destination as retrieved from the user's personal calendar, and suggest a route that is compatible with the user's stated preferences of not driving on highways whenever possible.

The above examples illustrate the different objectives of personalization for adapting interactions between persons and machines. These objectives include user interface customization, recording of user preferences and subsequent tailoring of delivered information items. Tailored information items include purchase recommendations, notification of events (for example, movie showings, registered reminders), news items from user specified interest categories (for example, sport game scores, stock quotes), and advertisements that relate to user shopping intentions. Not only can the information content be personalized, the rate of information updates can often be user selected. For example, on the My Yahoo! page, the user can select update time intervals from 15 minutes to several hours.

Personalization has also been applied to shared physical spaces where the physical environment, not just the virtual world, can adapt itself based on the context of the users currently present in the environment. Accenture Technology Labs implemented a shared space system in which played music can be adapted to the preferences of the majority of currently present users [7]. Such a system could be applied, for example, in a fitness center. Similarly, information displayed on a large monitor in a work environment's shared space can use knowledge about the preferences of employees that pass by to display information of common interest. Physical space adaptation requires a location determination capability. This capability could be realized by users carrying their mobile terminals, and the environment detecting user presence by detecting the terminals, or alternatively, the terminals could detect the fact that they moved in a particular space and provide a corresponding indication to the shared space system.

6.2 User models

User modeling has been used extensively in the design of user–computer interactions. The design of a user interface, the presented screens, the sequence of exchanges, all presuppose a type of user that will respond favorably to the computer generated questions and replies. This adaptation to user models is referred to as personalization, and can be built off one "typical" user model, or else a collection of such models when more refinement is desired in the exhibited computer's behavior. The users either specify these models explicitly or else the system infers them by observing user behavior.

In mobile environments, the user has a cell phone, or other mobile terminal that can be used to interact with an information system. For the information system to be more cooperative in data retrieval requests, it needs to draw assumptions about the user's beliefs, goals, and plans. This could be particularly challenging in mobile settings when no a priori user model is available. And even when there is a model, the specific circumstances of the user due to his or her mobility may affect the user's goals or plans. What may be relevant to a user in location *L1* at time *T1* may no longer hold when the user moved

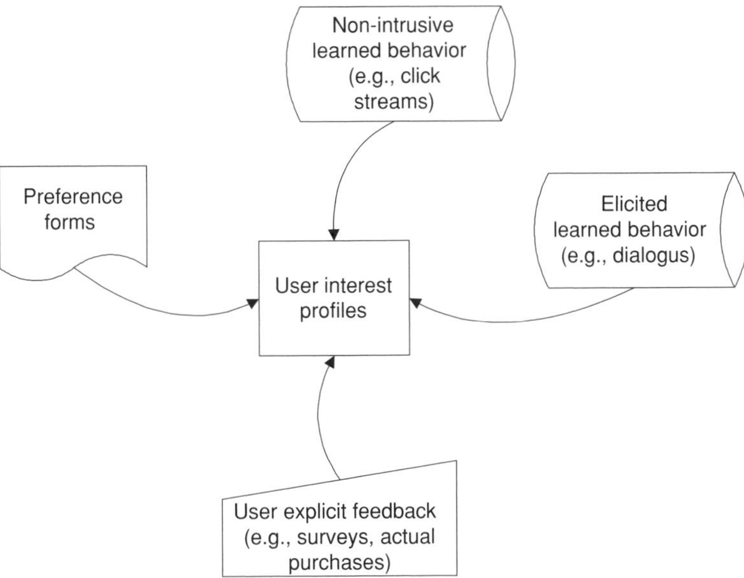

Figure 6.2 User model construction.

to location *L2* at time *T2*. In the following we expand upon the existing approaches for constructing user models and capturing the information in user interest profiles.

6.2.1 Explicit and learned behavior models

An explicit model specification requires the user to fill in preference forms where he can specify system parameters that will determine, for example, the interface screens that will be used whenever he interacts with the system (as in My Yahoo!). This approach to collect user profile information requires most effort on the part of the user (see Figure 6.2 for the different profile generation approaches). Furthermore, users may not always know how to best fill in the fields in their profile. For example, if information is requested about a user attribute where the user may have a wide range of preferences over time, then committing to one value may be over restrictive. A related issue is that a profile may become out-dated, unless the user puts in the extra effort to go and update it.

The alternative to explicit model building is for the system to learn user behavior through interactions, and to record these behaviors until, eventually, a personal user model is created. Learning about user behavior can be elicited through user dialogues. For example, an automated help system can adjust the response it provides to a user's question to the level of the question itself. This method assumes that the help system has a model of different user sophistication levels, and can map questions to different sophistication levels. This approach of user model building has the benefit of allowing fast adaptation to personal preferences; however, the adaptation is only to those user models that the system was designed to support.

Non-intrusive learned behavior is another approach for collecting user features in user model building. An example of this type of learning is click stream analysis, where user paths in site visits are tracked by recording link selections and time spent on each page ([8], [9], [10]). This approach has been used in ecommerce, where it was noticed that only a small percentage of visitors to retail websites will actually perform an online purchase. This has led some retail website owners to analyze click stream data to determine how users navigate through the website before making a purchase, tracking shoppers' steps backwards from checkout. The analysis results can lead to website redesigns that provide shoppers with the shortest route to what they're looking for.

A combined approach that leverages both learned behavior and explicit model building could overcome some of the problems associated with static profiles. Middleton *et al.* implemented a recommender that relies on user profiles for recommending research papers that are published on the Web [11]. The topics that are of interest to a user are recorded in the user's profile, and each topic has a numeric value that reflects the relative user's interest in the topic. The user profiles are updated dynamically by observing the papers that users browse. Whenever a paper is browsed that has a classified topic, an interest score for that topic is accumulated in the corresponding user's profile. In addition, the user can provide explicit feedback for each recommended paper in the form of "interested" or "not interested", and this feedback also accumulates interest values for topics. The interest topics for research papers are structured in a hierarchical ontology. Whenever the interest value for a topic in a user's profile is changed, the immediate super class receives 50 percent of the main topic's value. The next super class receives 25 percent, and so on, until the root of the tree is reached. This allows expansion of the list of topics that a user is interested in with more general topics that could be used in future recommendations. The resulting user profile is richer and more comprehensive.

Finally, traditional feedback survey forms, or recording of actual selections and purchases made by users, are yet another way of collecting user preference information. Typically, this data would be stored in what is termed long-term user interest profiles and enable businesses to build models that reflect the actions taken by their customer population. While electronic surveys may not always be applicable in a mobile setting, the unobtrusive recording of purchases done with a mobile terminal could certainly be applied as this would not affect the mobile user's current activity.

6.2.2 User stereotypes

Rich [18] has brought forth a concept of user stereotypes is to infer from a small number of facts a much larger set of user facts [12]. The idea is to quickly generate a user model from a small number of observed facts about the user. A stereotype is a collection of attribute-value pairs that attempt to characterize a user model. With each value is also associated a confidence in the actual value, as well as a list of reasons as to why the value is believed. As the result of an observed user's interaction with a system, a stereotype can be activated by the system. The stereotype, in turn, can predict values for other user attributes, and from these the system can know how to adapt its interaction with the user, and present relevant information to help the user accomplish tasks.

A book recommendation system, called Grundy [12], has implemented the concept of stereotypes. Grundy asks the user for a few keywords that provide a self-description that it uses to activate one or more stereotypes. If these are not sufficient to start giving book recommendations, then it may ask the user for a few more keywords. Once it has settled on a user model, it uses an inverted index into a book database to select all the books suggested by the attribute values of this model. Grundy tracks the book selection process and asks the user questions on why certain book selections are rejected. Based on the user's answers, it can update both the user's model and the database of stereotypes.

While a system such as Grundy could be used in a mobile setting for quickly generating a user model, a key issue that needs to be dealt with is how to ascertain the confidence level in the system's inference. The assumptions made by the system may be incomplete as some users may not fit into any of the system's stereotypes. In addition, the generated conclusions may be flawed; for example, if the user provides many keywords it may be more difficult to reach a recommendation that considers all the provided inputs since there may be fewer matches for a large set of keywords.

To answer the above possible misunderstandings between the user and the information system, the user could be presented with the assumptions that were derived about him or her. In some situations this may be prohibitive because of the large number of assumptions, or because of the nature of the mobile interaction that does not allow for extensive dialog. Alternatively, the information system could present its results with a confidence level whose value could warn the user about the results accuracy.

An interesting aspect of Rich's elaboration on user models is the separation between models of long-term user characteristics such as general interest areas, and short-term user characteristics such as the problem being currently solved. In the next chapter, we elaborate on the notion of user context and refer to these two models as static user context and dynamic user context, and show how they affect Web information adaptation.

6.2.3 Natural language interactions

Previously generated profiles that store user preferences are an often used source of information for discovering users' intents. The user profile could contain the requested information, for example, cuisine preferences if the user is seeking a restaurant. In other cases, a Web information service may need to gather more user attributes to decide what information to send to the user.

In mobile environments, the user will typically have access to terminals with limited input and output capabilities. Under these circumstances, the user may want to rely on natural language-based dialogs that could also be multi-modal, meaning that the mobile terminal will enable simultaneously both text and audio interactions. While natural language can be a strong enabler of personalization [13], computer understanding of natural language faces some difficult technical challenges. While understanding of predefined sets of keywords can be accomplished, the true difficulty lies in the contextual understanding of intentions. For example, if a user requests "to visit the main tourism attractions in a city", it may depend on the tourist's interests, the tourist's available time, on what is perceived as "main", and on the current accessibility of attractions. The system will have to take into account all these factors before suggesting a list of attractions.

Dialog systems attempt to circumvent some of the above difficulties by piecemeal collection of user inputs. A dialog management system decides what to ask the user, in what sequence, and also decides on the modality of the communication, that is whether information should be conveyed by voice, text, or graphics. The modality of communication, in particular, is yet another enabler of personalization since the interaction could revert to voice or text as the user situation changes.

User profiles can be associated with predefined user models, and the system can decide on a suitable dialog for eliciting more information based on a user's association with a given model [14]. For example, the choice of used terms and the level of difficulty of used sentences can be tailored to particular user models. Belkin [15], in particular, reports on how to suggest terms to a user for reformulating information searches. These personalized dialog interactions help drive the collection of lacking pieces of knowledge, so that ultimately the system is able to reach a better decision on the user's current needs and present suitable information.

6.3 Recommender systems

Recommender systems are as varied as help assistance programs that give the user suggestions on how to better use their service (for example, the Microsoft Word Office Assistant), to systems that help the user find information on a specific topic, to movie recommenders. While recommenders can be perceived as limiting the list of choices presented to a user, hence precluding a discovery experience, their objective is to support the fast retrieval of relevant information. With adequate recommenders, mobile users will not feel hampered by the fact that they are not in a familiar environment, may have limited available time, and can only rely on a terminal device with constrained input/output capabilities.

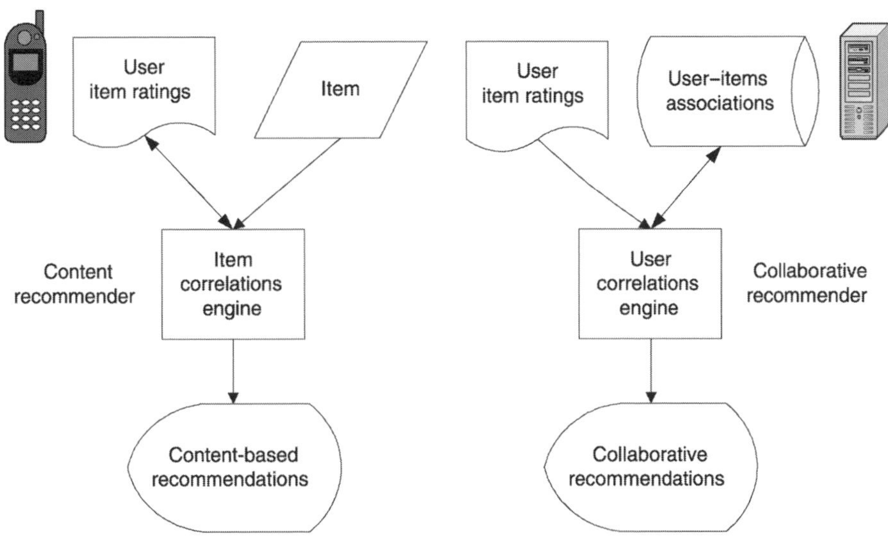

Figure 6.3 Content and collaborative recommenders.

6.3.1 Content and collaborative recommenders

Recommenders can be classified into two main groups: content-based and collaborative recommenders. Content-based recommenders rely on a user profile that records the user's history of past items accessed or purchased. For example, if a user has retrieved certain Web documents in the past, and has rated them, this information can be used to recommend new document items "similar" to those the user liked in the past. Since this approach relies on a private user profile, the profile can be stored in the user's mobile terminal and sent to the recommender upon request. The item correlations engine in Figure 6.3 is a content-based recommender that stores item ratings in the respective user profiles. When a recommendation is requested on a certain item for a given user, the recommender queries the user's profile and compares the item in question with previous user-provided item ratings.

Collaborative recommenders rely on information about a group of peer users that have similar likings to the user in question. Typically, a user will be asked to rate items (for example, books, movies, restaurants), and the recommender then matches the user to "neighbors" who have rated the items similarly. Information about the items these neighbors have purchased or used can be applied to come up with a recommended new item. Amazon uses a collaborative recommender to suggest new book purchases. In this system, the user can be exposed to new items that are different from those experienced in the past. This recommendation approach relies on neighbors' information that can span many users, and this information is typically stored on a network server. The user correlations engine in Figure 6.3 is a collaboration recommender that stores user–item associations in a database. When a recommendation is requested for a certain user, the recommender finds a group of similar users in the user–items associations' database. Based on this group, the recommender computes a weighted recommendation for the user.

Collaborative recommenders may not be very effective when the database of items that can be recommended is very large compared to the user population size, as many items may not be rated. Furthermore, newer items may not be rated if the items do change often. This is referred to as the rating sparsity problem. To alleviate this problem, the GroupLens document recommender from the University of Minnesota implements automated rating agents which evaluate new documents as soon as they are published and enter ratings for these documents [16]. Since each agent could have a dedicated algorithm for rating documents, these agents are treated just as other users that have their own rating preferences, and are considered part of the user population for generating a collaborative recommendation.

6.3.2 Hybrid recommenders

To improve the quality of recommendations, hybrid schemes have been proposed. For example, the Fab system, implemented at Stanford University, relies on user profiles that are based on content analysis, and these same profiles are used to determine groups of near users [7]. Fab will recommend Web pages to users when the pages score highly against their user profile and when they are rated highly by users with similar profiles. Another hybrid system is the MovieLens movie recommender website [17], from the University of Minnesota. MovieLens lets users specify a profile by giving examples of movies that they identify with a certain theme. It then recommends movies that match the theme and have been recommended by the user's neighbors, hence combining content and collaborative recommendations. The collaborative recommendation rank orders the content-based selections to improve the personalization of the presented list.

Other approaches at improving the operation of recommenders take into account additional elements of the user's context, for example, the user's location and time of day [18]. In addition, the recommender may prompt the user to provide specific inputs in the form of selected terms, or examples that help the recommender focus its search (see Figure 6.4). In most cases, the user will use a mobile terminal to provide the required data; in other instances, the user profile stored in a network server may provide the needed information. For example, past histories of user choices and actions may be stored on a network server.

When the user seeks a recommendation for an item for which there is no clearly defined query formulation, the user may have to engage in a series of interactions with the recommender until a satisfying response is obtained. For example, to specify a search, the user may have to iterate information a few times until he or she ends up using the right terms that the system can respond to. For example, the user may not

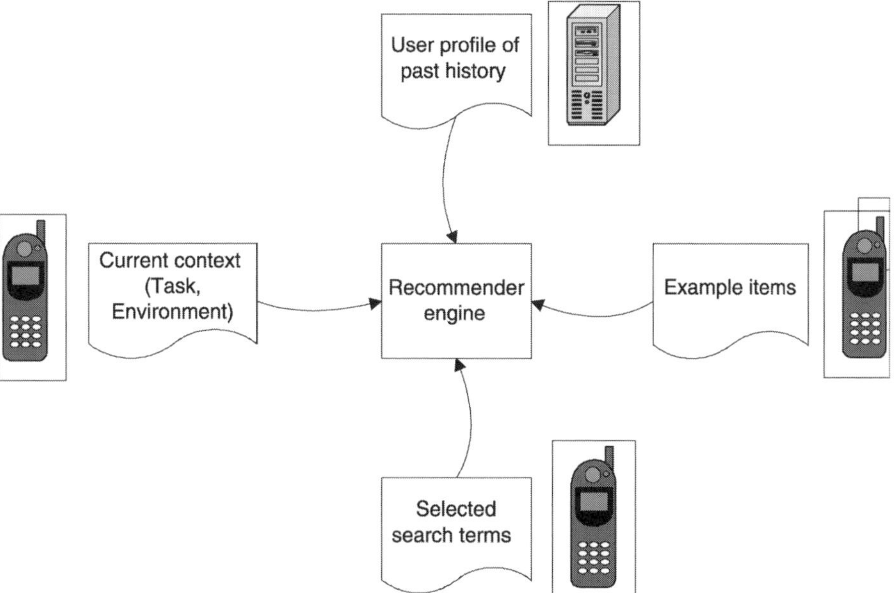

Figure 6.4 Data sources that focus the recommender's search.

know ahead of time what the best terms that describe his or her problem are or the terms used by the user may be different to those used by the information system in its internal representation. Research performed by Belkin at Rutgers University points out that "term-suggestion" recommenders are very efficient in the generation of search results [15]. Such recommenders propose search terms and engage the user in selecting relevant terms that can help the recommender reformulate its search queries until the right information is retrieved.

For many recommender systems, unlike the previous term-suggestion recommender, interaction with the user is limited and delivery of an information item to an online user is an activity triggered by the fresh availability of the item. Information delivery usually does not take into account the user's context, for example, the user's profile or the specific task that the user is currently attempting to accomplish. For example, classic movie recommendation systems will use only historical ratings data to come up with recommendations. Some recent recommendation systems consider also the user's context in the form of a task profile that consists of a list of example items that relate to the current recommendation request [17]. A user that wishes to purchase a gift for a child might specify to the recommender system two or three items that the child already owns and likes. Similarly, a user could specify movies that they saw and liked as examples of the kind of movie that they would want a recommender to suggest, as done in the MovieLens website recommender.

These profile-based recommenders are useful when users can easily describe examples of what they are looking for but have a hard time providing a formal description.

This is not unlike the term-suggestion system previously described, except that here the user is the one that provides the suggested terms. An interesting corollary of this is that the fewer examples the user provides, the more accurate the recommendation will be. While this may seem at first contradictory, it can be explained by the fact that the more examples the user provides, the more difficult it is for the recommender to come up with a result that matches all examples, and hence the lower accuracy.

6.3.3 User feedback in recommenders

Some recommenders include a mechanism by which they keep refining their recommendations as they collect user ratings on suggested items. The ability to record a user's specific selections enables the recommender to be personalized to a user's likings. A set of measurable attributes $\{a_i\}$ could be associated with each item in the system's database. Each user u has an associated list of weights $\{w_i{}^u\}$ that reflect the relative importance that he or she attaches to each attribute. This weight vector plays the role of a user model. The linear product of the attribute and weight vectors represents a utility function whose value is a relative cost that the user associates with a recommended item:

$$c_u = \sum_i \left(w_i^u \,.\, a_i\right)$$

The utility function enables the recommender to order the suggested items on the user's display. This latter ordering is of particular benefit to mobile users with small-screen terminals.

The recommender can adjust the user's weight vector with user feedback collected by the recording of online selections, or through electronic and mailed-in feedback forms. Unobtrusive feedback collection systems put less of a burden on the users; however, they do impact on the user's privacy. For example, a GPS enabled mobile terminal allows automated tracking of user paths, and, from this data, a navigation system can conjecture about the user preferences for certain routes. It is up to the user to check the recommender's privacy policy to verify how the collected data on user selections will be used.

A feedback collection system was implemented by Rogers and Langley [19] for mobile users that require driving route recommendations. Each recommended route has a set of measurable attributes that include the estimated total driving time, total distance, number of turns, and number of intersections. Their recommender provides a pair of suggested routes, and records the one selected by the user. The utility's function cost of the two routes is compared to the selection, and if they are not compatible, the user preference weights are adjusted to reflect the selection. An interesting observation is that user selections are not always according to the utility's

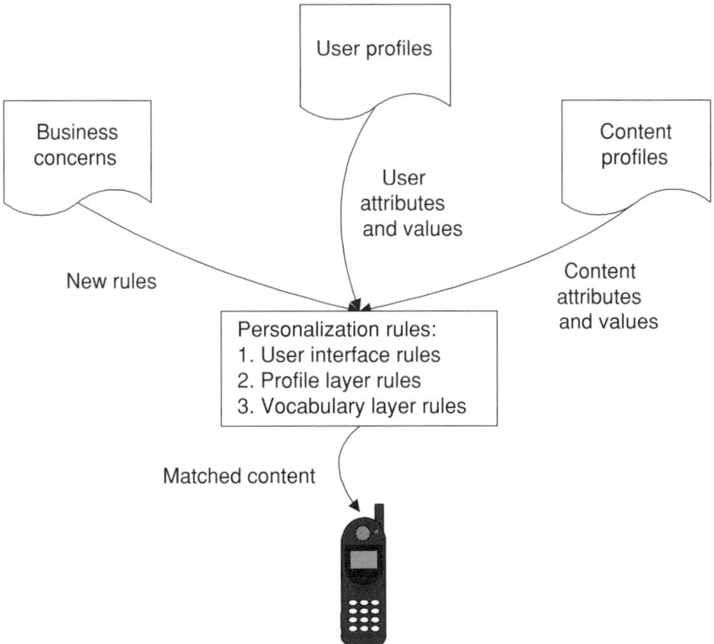

Figure 6.5 Abstract information personalization architecture.

function attributes. For example, users may use at times additional factors that are not represented as attributes, such as particular dislikes for certain routes or intersections. These cases represent a noise element in feedback collection, and it is up to the recommender designers to make sure that the corresponding user models are not affected.

6.4 Personalization architecture

An architecture that enables personalization needs to take into account the user profiles, the content served by Web servers, the business rules for serving content to mobile users, and matching mechanisms that can relate the searched-for items specified by the user with the system stored information. Such an abstract personalization architecture (and its associated data flows) was defined by Instone [20]. The architecture consists of three main components (see Figure 6.5):

- User profiles that represent user interests and behaviors; profiles have defined attributes and corresponding values.
- Content that is profiled by attributes and assigned values.
- Personalization rules that are derived from the business context of what to present and how to present it.

6.4.1 User profiles

Personal information can be collected from users explicitly, for example, when they fill in preference forms. Such a form was presented to a visitor to Motorola's wireless Web museum, and after specifying his or her fields of interest the visitor was presented with a relevant tour suggestion on a wireless terminal [21]. Alternatively and as previously described, personal information about the user can be learned from the user's behavior and collected unobtrusively. Web designers can embed scripts in Web pages to monitor user browsing actions that are then returned in cookies to the Web server. Alternatively, user actions can be derived from the analysis of Web server log files which record information like user name, host name, time and date, Web page requested, and any form values.

A user monitoring system was implemented by Cingil *et al.* [8] with a user agent on the client terminal tracking items such as user site visits, submitted queries, actions performed (for example, purchase, sell, view), accessed resources, and time spent with the resource. From this log of items, the system generates a user profile that can then be used for discovering Web content that is of interest to the user. User profiles can also be drawn on to find affiliations between users and generate corresponding groups in order for sites to make recommendations to users based on their group affiliation. The collected user preferences are then matched with website features so that a determination can be made on whether a site has relevant information for a visitor associated with a specific group.

6.4.2 Personalization rules

The set of attributes used to describe users or content, form a controlled vocabulary. The personalization rules will decide how to match up users and content after analysis of the supplied attribute values. Business concerns drive the specification of these rules. These concerns include goals such as targeted marketing, spot advertisement for mobile users (such as for advertising a current ongoing sale), enhancing the user online experience, and meeting user privacy concerns. This is not just a decision about the relevant content for a particular user, but also how, that is, in what format and when to present it. For example, if the personalization system is apprized of the user's intent to shop for a particular item, and the user happens to pass by a store that carries a related sale, then appropriate advertisements could be pushed to his or her mobile terminal.

Personalization rules can be partitioned into layers depending on the source of personal information and the affected area of personalization. Instone's information architecture provides a three-layer partition of the rules (see Figure 6.5). At the top, there are rules that decide on user interface issues such as screen layouts. Below, there are rules that deal with specific user profiles, and take into consideration the profiles' attribute

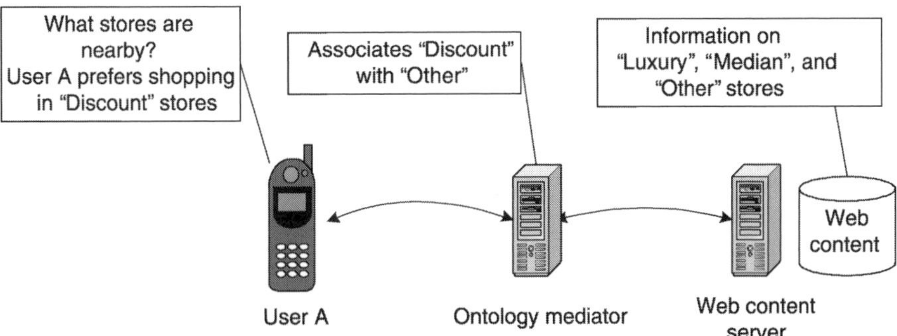

Figure 6.6 Ontology mediation of naming differences.

values to choose content that is relevant for display. Finally, at the bottom, there are rules that deal with the vocabularies used by the users and the content providers, and associate relationships between the terms, transforming the vocabulary into an ontology that can be reasoned with. For example, if a user is shopping for a tent, an inferencing rule can deduce that the user is interested in camping equipment, and related advertisements can be sent to his mobile terminal. These latter vocabulary rules are handled by what are termed ontology mediators.

6.4.3 Ontology mediators

The matching of users with content is simplified if the same terms are used to describe both user preferences and content features. Otherwise, a mediation function, referred to as an ontology mediator, is required to map between the terms (Campbell *et al.* ([22] and [23]) and Wiederhold *et al.* ([24] and [25]) define concepts for ontology mediation and composition). A mediator's service allows taxonomic distinctions to be overcome where, for example, the user's location can be categorized in the user profile as part of the user dynamic context, and in a Web content server it is part of the environment context. Similarly, context naming differences are mediated; for example, a user may request information about local stores, and his preference for "Discount" stores needs to be handled by a Web content server that provides data on local stores categorized as "Luxury", "Median", and "Other" (see Figure 6.6).

The matching operation between a user's profile and Web resources can take place either at the client side, at an intermediary trusted site, or at the source website. The chosen alternative depends on the client terminal capabilities, (that is, can it handle the processing, and the privacy constraints that dictate whether the client can trust the website with its user profile?. If privacy is not guaranteed on the server side, then the profile comparison can take place at the client, with the website providing a profile of its features to the client's terminal. If a trusted third party exists, it may gather the user profile and the website profile to discover interest overlaps. The W3C's

P3P protocol, described next, when implemented by both client and server, can help determine any privacy restrictions in the interactions between these parties.

6.5 Privacy concerns and the P3P initiative

The effective operation of mobile recommenders often relies on the existence of user profiles. As previously described, content-based recommenders rely on user profiles of item ratings. While a profile could be stored on the user's mobile terminal, other recommender schemes, for example the Fab hybrid recommender [7] requires multiple user profiles, typically stored in the network, when determining a group of neighbors for collaborative filtering. This reliance on user profiles may be of concern to users that are apprehensive of how information in their profiles is being used. Businesses are realizing that to lure customers they must provide both personalization and a promise that the personal information they gather will stay confidential. The collection and use of private information is often regulated by legislation. In the USA, individual states consider every year many bills that pertain to privacy. Privacy laws put constraints on the use of personal information and often require companies to receive the consent of their customers before sharing information with third parties. Similarly, in the UK, the Data Protection Act 1998 legislates on the use of personal data by companies and individuals.

6.5.1 W3C's Platform for Privacy Preferences (P3P)

To alleviate some of the user privacy concerns, the W3C started the Platform for Privacy Preferences (P3P) initiative to determine an overall architecture for enabling privacy on the Web for browser-based clients [26]. The intent of P3P is to provide for automatic determination of what personal information can be shared with a website from a comparison of a website's privacy policy with the user's privacy preferences. Although P3P specifies privacy policies, it does not enforce them. Enforcement can be applied by third party organizations that audit businesses to ensure that they are indeed following their stated policies. TRUSTe [27] is such a watchdog organization that performs ongoing monitoring of the privacy practices of websites that have registered with them. The registered websites display a TRUSTe logo that indicates compliance with their stated policy.

P3P-enabled websites make their privacy information available in a standard, machine-readable format, as well as in human-readable text. A user agent such as a P3P-enabled browser, or a plug-in, can interpret this information automatically and compare it to the user's own set of privacy preferences. If the site's privacy policy meets the user's preferences then the user interaction with the site can proceed with no interruptions; otherwise, the user agent could present appropriate warnings to the user.

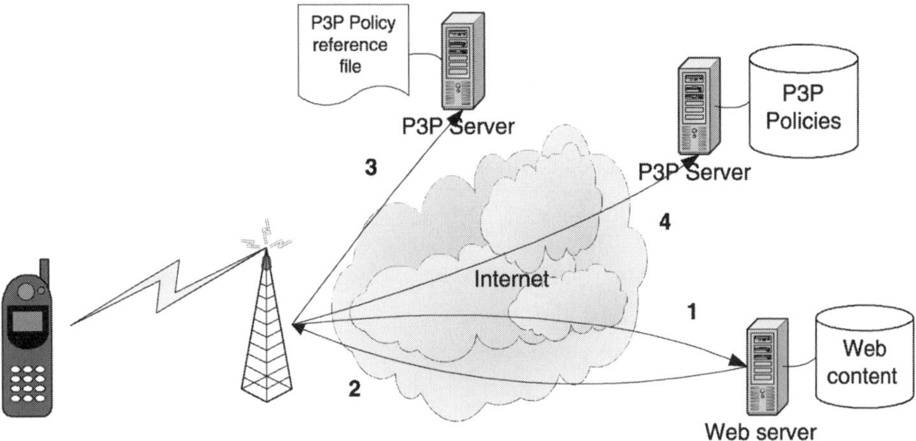

Figure 6.7 P3P privacy policy check steps.

The comparison of a site's policy with user preferences could alternatively be implemented in a proxy server. The proxy server would retrieve user privacy preferences from the user's mobile terminal, or from a website that stores user profiles. In addition, the proxy server would retrieve the website's privacy policy. In this latter case, the user relies on the trustworthiness of the proxy server for evaluating the privacy risks that he or she may encounter when interacting with a site.

6.5.2 Locating P3P policies

As a privacy policy can be uniquely associated with a URI, it often suffices to receive a policy's URI to determine a website's policy. Furthermore, rather than transferring policy files, it is more economical to retrieve URI references to P3P policies. Policy reference files store URIs where P3P policies can be found. To process the policy that applies to a given resource, one needs first to locate the policy reference file for that resource. After fetching the policy reference file, the relevant URI can be extracted. If the detailed policy is still required, the policy document can be retrieved by following this latter URI.

The policy reference file can be located in a well-known site. Alternatively, it can be retrieved from an included URI in a P3P HTTP header, or it can be reached via a link embedded in received markup. The steps for processing a privacy policy check are shown in Figure 6.7 where the policy check is performed by the mobile terminal and the URI to a policy reference file is retrieved from a returned P3P HTTP header. Initially, the mobile terminal issues a resource request to a website (step 1); the website returns the URI to a policy reference file in a P3P HTTP header (step 2). The mobile terminal requests the policy reference file at the provided URI (step 3); upon receiving the reference file, the mobile terminal can either determine the site's privacy policy from the resource policy file's URI, or else it requests the resource policy file at this URI (step 4). Upon receiving the policy file, the mobile terminal can parse the file's contents and determine if the website's policy meets the user's privacy requirements.

Separate P3P policies could apply to different parts of a website, where each part is uniquely identified by its associated URI. For example, in an ecommerce site, one policy could apply to catalog shopping, where the site tracks the items purchased by the user, while another policy could apply to checkout, where the site collects the user's address and credit card information.

Different user types could also be subject to different policies of a website, depending on the services offered by the website. An ecommerce website's privacy policy for registered users could spell out how the site intends to use user information. For example, registered users could receive periodic emails on advertised products, based on their stated product interests when they registered. The website could have a separate privacy policy for non-registered users. The website could track the selections made by these users when they browse the website's pages, and offer them instantaneous suggestions for product purchases.

6.5.3 P3P policy representation

A P3P policy is a website's XML representation of its privacy practices. P3P policies include information about the entity publishing the policy, information about the location of a human-readable policy, information about how individuals can access their own collected data, and information about how individuals can resolve privacy-related disputes with the entity. In addition, a P3P policy describes specific data practices including the type of data collected, the purposes for which it may be used, the types of recipients with which it may be shared, and the applicable retention policy.

Listing 6.1 shows an example policy from the W3C's P3P specification [26]. This policy, coded in XML, is of an example company, CatalogExample, and expresses a privacy policy for users that browse the CatalogExample website. In this case, the collected information on users is used to improve the site, not to identify the users for other purposes. P3P policies can also be coded in the Resource Description Framework Schema (RDFS) language defined by the W3C for representing resources. A description of how to use RDFS for that purpose is provided in [28].

```
<POLICIES xmlns="http://www.w3.org/2002/01/P3Pv1">
  <POLICY name="forBrowsers"
    discuri="http://www.catalog.example.com/
    PrivacyPracticeBrowsing.html"
```

```
xml:lang="en">
<ENTITY>
  <DATA-GROUP>
    <DATA ref="#business.name">CatalogExample</DATA>
    <DATA ref="#business.contact-info.postal.street">
     4000 Lincoln Ave.</DATA>
    <DATA ref="#business.contact-info.postal.city">
     Birmingham</DATA>
    <DATA ref="#business.contact-info.postal.
     stateprov">MI</DATA>
    <DATA ref="#business.contact-info.postal.
     postalcode">48009</DATA>
    <DATA ref="#business.contact-info.postal.
     country">USA</DATA>
    <DATA ref="#business.contact-info.online.email">
     catalog@example.com</DATA>
    <DATA ref="#business.contact-info.telecom.
     telephone.intcode">1</DATA>
    <DATA ref="#business.contact-info.telecom.
     telephone.loccode">248</DATA>
    <DATA ref="#business.contact-info.telecom.
     telephone.number">3926753</DATA>
  </DATA-GROUP>
</ENTITY>
<ACCESS><nonident/></ACCESS>
<DISPUTES-GROUP>
    <DISPUTES resolution-type="independent"
     service="http://www.PrivacySeal.example.org"
     short-description="PrivacySeal.example.org">
    <IMG src="http://www.PrivacySeal.example.org/
     Logo.gif" alt="PrivacySeal's logo"/>
    <REMEDIES><correct/></REMEDIES>
  </DISPUTES>
</DISPUTES-GROUP>
<STATEMENT>
  <PURPOSE><admin/><develop/></PURPOSE>
  <RECIPIENT><ours/></RECIPIENT>
  <RETENTION><stated-purpose/></RETENTION>
  <DATA-GROUP>
    <DATA ref="#dynamic.clickstream"/>
```

```
            <DATA ref="#dynamic.http"/>
          </DATA-GROUP>
        </STATEMENT>
      </POLICY>
    </POLICIES>
```

Listing 6.1 P3P privacy policy of the CatalogExamples website

In the CatalogExample policy file from Listing 6.1, a *<POLICIES>* element can contain multiple *<POLICY>* elements, one for each defined website policy. Each *<POLICY>* element defines the specifics of a policy, and includes a mandatory *discuri* attribute that contains the URI of the human-readable privacy statement. The *<ENTITY>* element gives a description of the legal entity making the representation of the website's privacy practices. The following *<ACCESS>* element indicates if the user can view identification data, that is personal contact and account information; in the above example, the website specifies that it does not collect identification data.

The *<DISPUTES>* element describes dispute resolution procedures for disputes about a website's privacy practices. In the above example, the user can complain to an independent organization identified by the name in the *short-description* attribute, and reachable at the URI described by the *service* attribute. The enclosed *<IMG>* element contains an image logo of the independent organization, and the *<REMEDIES>* element specifies the remedies in case of a policy breach. In the above example, the website will remedy any error or wrongful action arising in connection with the privacy policy.

The *<STATEMENT>* element specifies how the data referenced by the enclosed *<DATA-GROUP>* is handled. The *<PURPOSE>* element specifies how the collected data will be used. In the above example policy, user data is collected for technical support and development of the website and its computer system. The *<RECIPIENT>* element specifies who will be receiving the collected data; just the website's organization in this example. The length of time that the collected information is retained is specified in the *<RETENTION>* element; it is retained here just for the stated purpose, and will be discarded at the earliest time possible.

Finally, the specific data that is collected by the website is described in the *<DATA>* elements within *<DATA-GROUP>*. The description is based on the P3P defined data schema that enables websites to communicate to user agents about the data that they collect. In this example, the website collects two types of dynamic data: click-stream and HTTP protocol information. Click-stream data is typically found in webserver logs and includes information such as IP address or hostname of the requesting client, the URI of the requested resource, the time the request was made, the HTTP method in the request, and the HTTP status code in the response.

As can be seen in the above policy file, policies may contain many items of information. The use of URIs to identify policies is more economical, particularly in wireless environments where the bandwidth is more limited and mobile terminals have less computational power to process policies. Listing 6.2 of a policy reference file from the P3P specification shows how a website's P3P policies are associated with URIs.

```
<META xmlns="http://www.w3.org/2002/01/P3Pv1">
 <POLICY-REFERENCES>
  <EXPIRY max-age="172800"/>
  <POLICY-REF about="/P3P/Policies.xml#first">
   <INCLUDE>/docs/*</INCLUDE>
   <EXCLUDE>/docs/finance/*</EXCLUDE>
   <METHOD>GET</METHOD>
   <METHOD>HEAD</METHOD>
  </POLICY-REF>
  <POLICY-REF about="/P3P/Policies.xml#second">
   <INCLUDE>/docs/*</INCLUDE>
   <EXCLUDE>/docs/finance/*</EXCLUDE>
   <METHOD>PUT</METHOD>
   <METHOD>DELETE</METHOD>
  </POLICY-REF>
 </POLICY-REFERENCES>
</META>
```

Listing 6.2 P3P policy reference file

The complete policy reference file is contained in the *<META>* element. The *<POLICY-REFERENCES>* element can contain multiple policy references, each in a separate *<POLICY-REF>* element. The *<EXPIRY>* element specifies the lifetime of the reference file, and applies therefore to all policy references. Each *<POLICY-REF>* specifies in the *about* attribute the URI where the policy can be found. The *<INCLUDE>* and *<EXCLUDE>* elements specify that the policy applies to the entire website's URIs that match any of the *<INCLUDE>* elements, but does not apply to those URIs that match the *<EXCLUDE>* elements. In this example, the two listed policies do not apply to any of the resources under the */docs/finance/* directory.

Each *<METHOD>* element in the above policy reference file is used to state that the enclosing policy reference only applies when the specified method is used to access the referenced resources. The first policy reference, */P3P/Policies.xml#first*, applies therefore to all resources whose path begins with /docs, but excludes those whose path begins with /docs/finance/, for the HTTP requests GET and HEAD. The second policy reference, */P3P/Policies.xml#second*, applies to the same resources, but only when the resources are accessed with the HTTP requests PUT and DELETE.

REFERENCES AND FURTHER READING

[1] Personalization Consortium, http://personalization.org/.

[2] My Yahoo!, http://my.yahoo.com/?myHome.

[3] U. Manber *et al.*, Experience with personalization on Yahoo! *Comm. of the ACM*, **43**:8 (Aug. 2000), 35–9.

[4] Staples, http://www.staples.com/.

[5] Amazon, http://www.amazon.com.

[6] J. Riedl, Personalization and privacy. *IEEE Internet Computing* (Nov.–Dec. 2001), 29–31.

[7] M. Balabanovic and Y. Shoham, Content-based, collaborative recommendation. *Comm. of the ACM*, **40**:3 (Mar. 1997), 66–72.

[8] I. Cingil *et al.*, A broader approach to personalization. *Comm. of the ACM*, **43**:8 (Aug. 2000), 136–41.

[9] M. Claypool *et al.*, Inferring user interest. *IEEE Internet Computing* (Nov.–Dec. 2001), 32–9.

[10] R. Whiting, Turning browsers into buyers. *Information Week* (Jul. 31, 2002).

[11] S. E. Middleton *et al.*, Capturing knowledge of user preferences: Ontologies in recommender systems. *Ist Int. Conf. on Knowledge Capture*, Oct. 2001.

[12] E. Rich, Users are individuals: Individualizing user models. *Int. J. Human–Computer Studies*, **51** (1999), 323–38.

[13] W. Zadrozny *et al.*, Natural language dialog for personalized interaction. *Comm. of the ACM*, **43**:8 (Aug. 2000).

[14] A. Kobsa, User modeling in dialog systems: Potentials and hazards. *AI and Society: The Journal of Human and Machine Intelligence*, **4**:3 (1990), 214–40.

[15] N. J. Belkin, Helping people find what they don't know. *Comm. of the ACM*, **43**:8 (Aug. 2000), 59–61.

[16] B. M. Sarwar *et al.*, Using filtering agents to improve prediction quality in the GroupLens research collaborative filtering system. *ACM CSCW 98* (1998).

[17] J. L. Herlocker and J. A. Konstan, Content-independent task-focused recommendation. *IEEE Internet Computing* (Nov.–Dec. 2001), 40–7.

[18] A. Pashtan *et al.*, CATIS; A context-aware tourist information system. *Proc. of the 4th Int. Workshop of Mobile Computing*, Rostock (Jun. 2003).

[19] S. Rogers and P. Langley, Personalized driving route recommendations. *Proc. of AAAI Workshop on Recommender Systems* (1998), 96–100.

[20] K. Instone, *Information Architecture and Personalization*. Argus Associates, http://argus-acia.com/argus_content/index.html (Dec. 2000).

[21] A. Pashtan *et al.*, Adapting content for wireless web services. *IEEE Internet Computing* (Sep.–Oct. 2003).

[22] A. E. Campbell *et al., Ontological Mediation: An Analysis*, Dept. of Comp. Sc., State University of New York at Buffalo. Feb. 1, 1995.

[23] A. E. Campbell and S. C. Shapiro, Ontological mediation: An overview. *Proc. of IJCAI Workshop on Basic Ontological Issues in Knowledge Sharing*. (Menlo Park, CA: AAAI Press, 1995).

[24] G. Wiederhold and M. Genesereth, The conceptual model for mediation services. *IEEE Expert* (Sep.–Oct. 1997), 38–47.

[25] G. Wiederhold and J. Jannink, Composing diverse ontologies. *8th Working Conference on Database Semantics (DS-8)*, Rotorua, New Zealand (1999).

[26] M. Marchiori *et al., The Platform for Privacy Preferences 1.0 (P3P1.0) Specification.* W3C Recommendation (Apr. 16, 2002), http://www.w3.org/TR/P3P/.

[27] TRUSTe, http://truste.org.

[28] B. McBride *et al., An RDF Schema for P3P.* W3C Note (Jan. 25, 2002), http://www.w3.org/TR/p3p-rdfschema/.

J. M. McCarthy, The virtual world gets physical: Perspectives on personalization. *IEEE Internet Computing* (Nov.–Dec. 2001), 48–53.

7 Ontologies and RDF Schema

Mobile terminals and mobile services are an integral part of the extended Web that includes the wireless domain, and as such require descriptions that facilitate automated interoperation between terminals and network servers. This chapter starts by reviewing ontology concepts and their application to enable the wireless Semantic Web. We review the W3C CC/PP framework for sending profiles of wireless terminals so that content servers can adapt their content to terminal characteristics. The FIPA agent standards organization's reliance on ontologies for mobile services is reviewed next. We end the section on ontologies by describing approaches taken to generate ontologies and report on commonly accepted criteria used to design and evaluate ontologies.

The W3C-defined RDF and RDF Schema (RDFS) are introduced next. RDF and RDFS provide a formal framework for defining an ontology, and are used in chapter 8 to describe an ontology of mobile user context. We review the concept of RDF graph models used to represent resources, their properties and associated values, and follow with a description of the RDF XML syntax with examples that clarify RDF statements. RDFS adds typing facilities that enhance RDF and enable the description of basic ontologies. Finally, we conclude this chapter with a review of Web ontology languages that build on RDFS to provide more powerful expressions of resource properties and resource relationships.

7.1 Ontologies for mobile services

7.1.1 Concepts and use

Ontologies were developed in the artificial intelligence community for formal knowledge representations of concepts and their relationships. An often-cited definition is the one provided by Gruber: "An ontology is an explicit specification of a conceptualization" [1]. An ontology's formal specification consists of the list of terms used in the domain of discourse and the relationships between these terms. The conceptualization is not concerned with the assignment of meaning to terms, but rather with a formal structure of reality as perceived by the end users of the ontology. This latter structure is valid across different situations that are applicable to the dealt-with reality. For example, the conceptualization of mobile phone devices should not be dependent on particular implementation instances. Campbell and Shapiro [2] have further refined Gruber's definition to include a set of axioms that constrain interpretation and well-formed use of the ontology terms.

How is an ontology used once it is defined? Gruninger and Lee [3] have stated three main reasons for using ontologies:

- For communication purposes to facilitate the exchange of information between cooperating parties by enabling a shared understanding about a domain.
- For drawing inferences on the internal structure and operation of an implemented system.
- For the reuse of domain information. Intelligent agents can extract and aggregate information from multiple parties to provide a comprehensive view of a domain's knowledge.

Examples of developed ontologies include RosettaNet [4], an industry consortium which defined a language for ebusiness processes and messages between supply chain partners for the purpose of establishing efficient trading networks, the Systematized Nomenclature of Medicine (SNOMED) [5], a large medicine vocabulary that enables clinicians, researchers and patients to share health care knowledge, and WordNet [6], a lexical database where English nouns, verbs, adjectives and adverbs are organized into synonym sets. In chapter 8 we provide a detailed ontology of a mobile user's context, and show how the various terms are logically organized, and how sets of terms can be used to determine modes of mobile service.

In the Berners-Lee vision of the Semantic Web [7], computers will be able to reason about Web data and leverage the right sources of data for the tasks they are attempting to perform. This requires the ready availability and acceptance of ontologies about the data. For example, search operations done by humans sifting through Web pages will be replaced by intelligent agents that can detect ontology pages linked to information pages, and reason about the information with the help of the related ontology. Heflin and Hendler described how page annotations can improve Web searches [8]. Similarly, a scheduling agent and a service provider can negotiate an appointment date using shared ontologies about schedules and services.

So the future of the Semantic Web is very much dependent on the future of ontologies. Ontologies can be standardized so that different entities can be preprogrammed with the understanding of the features embedded in the ontology. The standardization process is usually lengthy and requires agreement by many stakeholders. A more flexible approach is one where communicating agents exchange ontologies required for the discussion. This would usually require mediating services that can take the ontology of a requestor and reformulate requests using the ontology of the service provider. This is similar to using the services of a human translator in meetings between representatives of different countries, where the translator reformulates questions from one language to another.

7.1.2 Ontology exchange

The ability to exchange ontologies is very useful in dynamic environments where not all the elements are known ahead of time. Campbell and Shapiro ([2] and [9]) introduced the concept of an ontological mediator that enables communication between agents that use different ontologies. Similar concepts were defined by Wiederhold *et al.* ([10] and [11]). Ontological mediators are aware of the meaning of the exchanged messages and are able to bridge between disparate agents by translating between their differing taxonomies. In their model, every pair of communicating agents has its own ontological mediator that knows something about their respective ontologies. The mediator uses this meta-ontological knowledge about the terms used by the agents to find which expressions are used to denote the objects dealt with.

Using a heuristic graph traversal algorithm, the mediator can traverse the respective ontological trees to relate terms between ontologies. By checking superclasses, subclasses, and siblings of terms, the mediator can try to find correct translations. For example, if a mobile user's agent requests the location of discount stores the mediator may have to map the concept "Discount" to the ontology of a shopping service that provides store locations. The shopping service may use a different term, for example, "Inexpensive" to denote the same concept.

Alternatively, a mediator can ask an agent for questions to help place a term of one agent in the ontology of the other. The questions are essentially a learning approach to determine the context of a term's use from which its ontology placement can be asserted. This method is useful for classifying new terms that do not exist in one of the ontologies. For example, in the same example from above, the shopping service may not have a category of "Inexpensive" stores, however it has defined categories of "Luxury", "Median", and "Other". The ontology mediator may, following a questions and answers session with the service, determine to list stores from the "Other" category when it receives a query for "Discount" stores. In this case, the taxonomies used by the user and service are different as can be seen from the different concept trees, and the terms are different as well (Figure 7.1).

7.1.3 W3C Composite Capabilities/Preference Profiles

W3C's Composite Capabilities/Preference Profiles (CC/PP) provides an environment where cell phones can send a profile of their supported features (for example, screen size) to a Web service. The profile, referred to as a "user agent profile", consists of an ontology and values associated with the terms of the ontology. Based on this information, Web

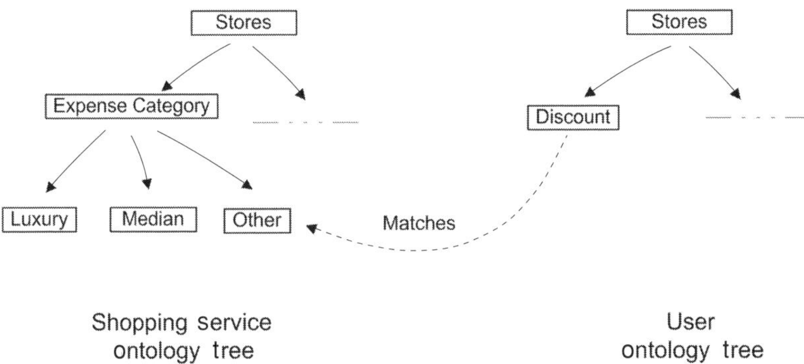

Figure 7.1 Service and user ontology trees: matching terms.

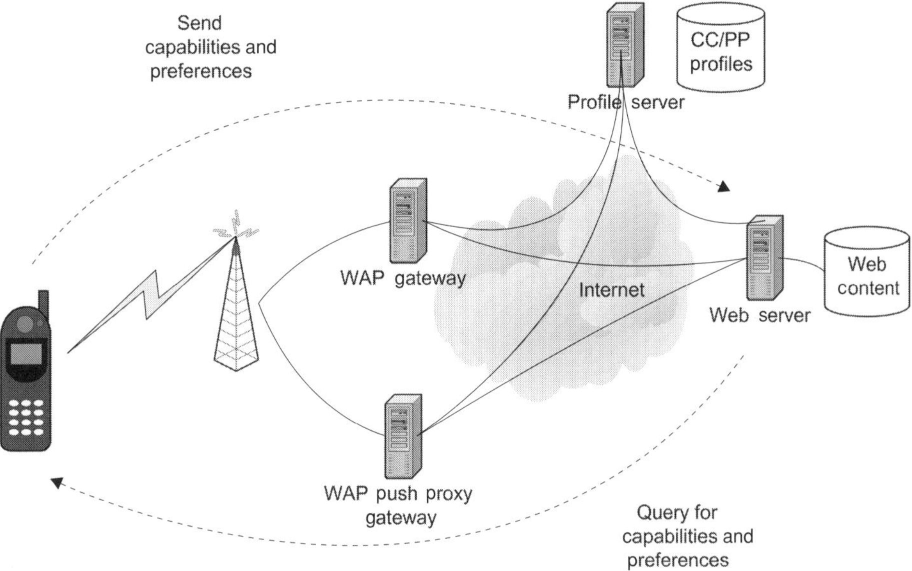

Figure 7.2 CC/PP architecture in a WAP environment.

content sent by the Web service to the cell phone can be adapted dynamically. Other context information can be sent with the same CC/PP capabilities. For example, a mobile user can venture in a new environment that is instrumented with special sensors, for example, to measure noise and light levels. A profile of these sensors can be sent to a Web service so that additional sensor-measured context information is taken into account for Web content adaptation.

The architecture in Figure 7.2 describes how a client's mobile terminal can send its user agent profile to the Web content server in a WAP environment. When the client initiates the request for information, the terminal sends the profile to the Web server via the WAP gateway. The profile can consist of a URL that points into a repository of user agent profiles. In this case, either the WAP gateway can resolve the reference or else

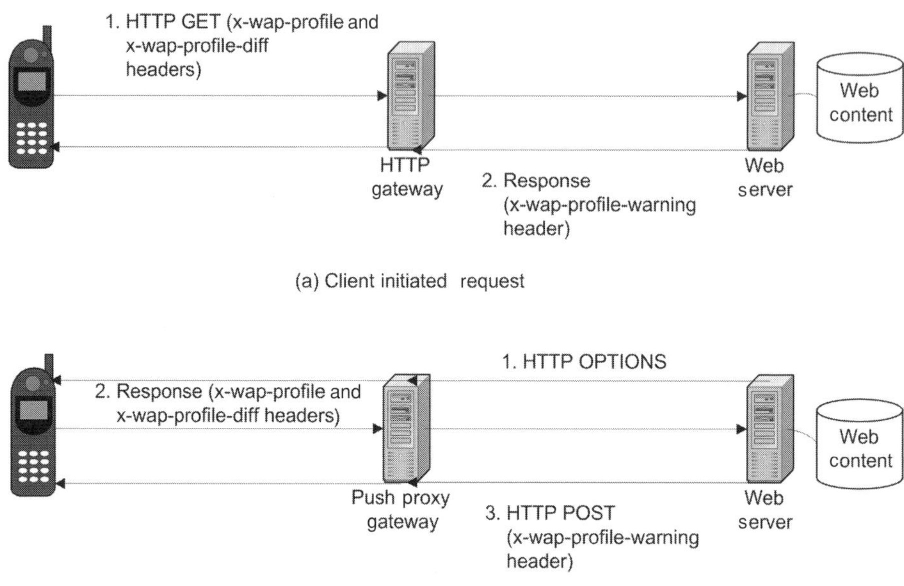

Figure 7.3 Wireless Profiled HTTP protocol for client profile transport.

the Web server can do so. The wireless network operator can, in the WAP gateway, add information to the retrieved profile that is not available to the client's mobile terminal. For example, the added information can be extracted from a Home Location Register (HLR) that contains user account information, or else the network operator may add information that reflects policies set by the operator, such as priority of content type delivery.

In the case where the Web server initiates the sending of information to the client's terminal, then the Web server requests client profile information from the Push Proxy Gateway (PPG). The PPG, in turn, requests this information from the client's terminal and can access the profiles repository to retrieve a profile as shown in Figure 7.2. The client's profile is then returned to the Web server for content adaptation prior to issuing the content push. As in the client initiated content request case, the PPG can also augment the client's profile with new information deemed relevant by the network operator.

The transport of profiles from client to origin servers can take two forms. If the client's terminal supports HTTP over-the-air then the wireless profiled HTTP protocol is used to relay the client profile information. See Figure 7.3 for a message exchange sequence that will result in the delivery of a client's profile to a content server. This protocol provides for extension headers in the HTTP message. There are two such headers; the *x-wap-profile* header that contains a URI referencing the client's terminal profile, and the *x-wap-profile-diff* header that contains profile differences in the form

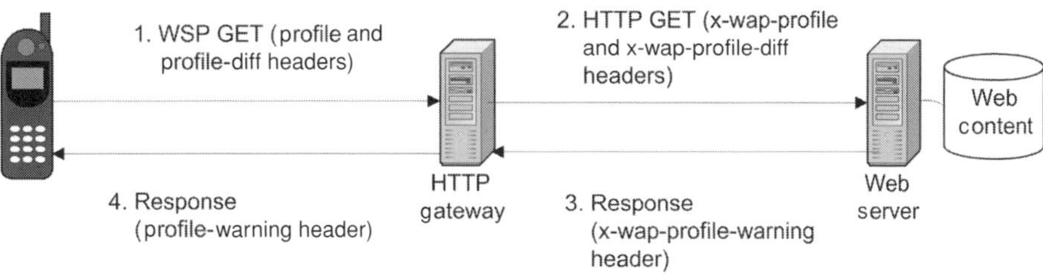

(a) Client initiated request

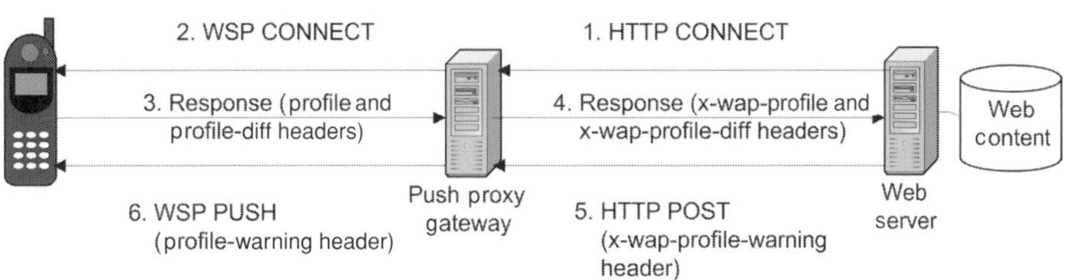

(b) Server initiated push

Figure 7.4 WSP protocol for client profile transport.

of XML documents. These latter XML documents are merged with the client's profile referred to by the first header. In the case of a server initiated content push, the client's terminal is first queried for its capabilities with the HTTP OPTIONS message. The client's terminal response message includes the *x-wap-profile* and *x-wap-profile-diff* headers.

If the client's mobile terminal uses the WAP protocols for connecting with the network infrastructure, then a CC/PP Exchange Protocol (CC/PPEX) is used over the wireless session protocol (WSP). Figure 7.4 shows the corresponding message exchange for the delivery of a client's profile to a content server. Two WSP headers, the *Profile* and the *Profile-Diff*, are used over the air to convey client terminal profile information in a similar way as in the wireless-profiled HTTP case. Upon receiving these headers, the WAP gateway generates corresponding *x-wap-profile* and *x-wap-profile-diff* HTTP headers that are sent to the origin server. In the case of a server initiated content push, the server request for a connection is converted to an over-the-air WSP connect request. Upon receiving the latter message, the client conveys its profile information to the PPG in the same WSP headers as in the client initiated request. These WSP headers are then converted to HTTP headers that are returned to the content server.

After adapting the server content according to the client's terminal profile, the Web server returns the requested content with a warning header, *x-wap-profile-warning*, that indicates whether the adaptation was fulfilled, partially fulfilled, or could not be done. The warning header is translated into a matching *Profile-Warning* WSP header for over-the-air transmission when the connection to the client's terminal is over the WSP protocol.

CC/PP provides a vocabulary that consists of a standardized set of attributes expressed in the resource description framework (RDF) format for describing user agent profiles that convey client capabilities and preferences. The CC/PP vocabulary defines a small, core set of features that are applicable to a wide range of user agents. Each vocabulary term is in the form of a URI that combines an XML namespace URI with a local XML element name. For example, the namespace URI *http://w3c.org/ccpp-core-vocabulary/* is combined with the element name *type* to yield the term URI *http://w3c.org/ccpp-core-vocabulary/type*. This core set can be extended with specialized vocabulary extensions that apply to different domains. Interoperability between clients and servers is enabled when they use the same extension vocabularies. For example, the WAP Forum has defined an extension vocabulary, referred to as UAProf, for describing the capabilities of cell phones.

7.1.4 FIPA's ontology service

The Foundation for Intelligent Physical Agents (FIPA) has recognized the need for shared ontologies between communicating agents that wish to work on common tasks [12], [13]. The shared ontology will provide both a common application vocabulary and a shared meaning of the used terms. Such ontologies should preferably be explicit, meaning that they have a declarative representation and are managed separately from the client applications. A dedicated "ontology agent" can manage such an explicit ontology, for example a device ontology [14], and control access requests from communicating agents.

An ontology agent can play a mediator's role and disambiguate terms using ontology structures, relate between ontologies, or translate terms from one ontology to another. For example, a querying client in Figure 7.5 can ask for the available "Quality" of service when the ontology O1 of the service does not include the term "Quality". In this case, the ontology agent can disambiguate the term "Quality" with the help of ontology O2, and provide the user with the two options "Best" and "Normal". In another example, the ontology agent could help two user terminals exchange pictures, when the capabilities of these terminals are expressed in profiles that rely on different ontologies shown in Figure 7.5, with user A's profile expressed in ontology O1 and user B's profile expressed in ontology O2. User A could query the ontology agent for the capabilities of user B's terminal. The ontology agent can help determine that user B's terminal does not support color and has a smaller screen than user A's terminal, so that a color picture sent from user A needs to be converted to grayscale and reduced in size before being sent out.

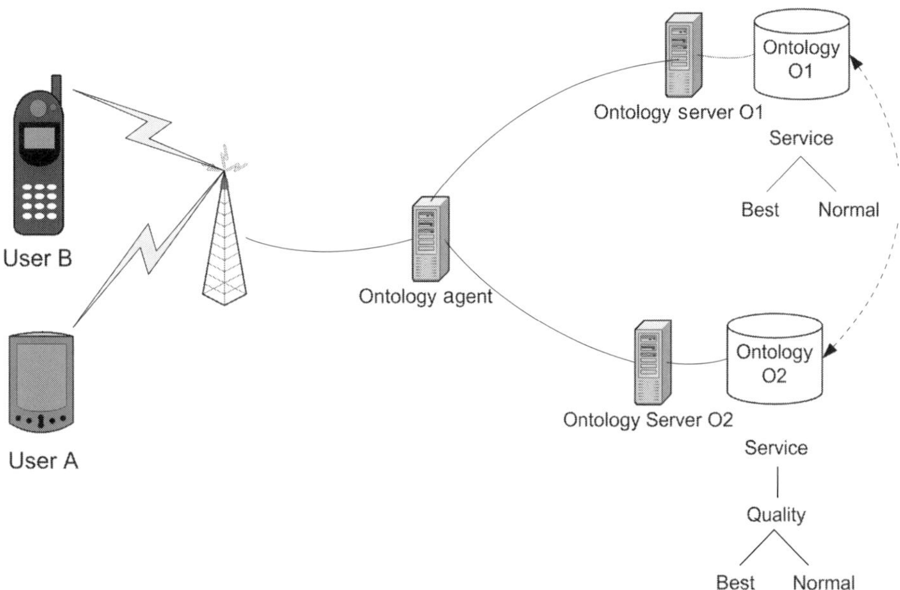

Figure 7.5 FIPA's Ontology Agent mediation function.

FIPA has generated application specifications that provide reference architectures for multi-agent systems with associated content ontologies. For example, the Personal Travel Assistance (PTA) specification [15] describes a multi-agent system for making travel arrangements taking into account travel preferences stored in user profiles. Different agents collaborate for travel planning and monitoring; these agents include a personal travel agent that acts on the user's behalf and is aware of the user's preferences, a broker agent for finding specialized travel service agents, and the travel service agents that can handle flight plans, hotel arrangements, car reservations, restaurant information, etc. A shared ontology can enable a common understanding and collaboration between these different agents (see Figure 7.6). The PTA specification provides such an initial ontology of trips and travel preferences.

7.1.5 Ontology design

How is an ontology generated?

Generating an ontology is a creative process that can be done by an individual researcher that analyzes a domain of interest and determines the relevant terms and relations based on his or her observations. Alternatively, ontology definition can be a collaborative effort done by a group of organizations that iteratively reach a consensus on the specification. The latter approach can be challenging and time consuming since it requires consensus between multiple cooperating parties that may have different views, different assumptions, and different cultural backgrounds. Collaborative efforts

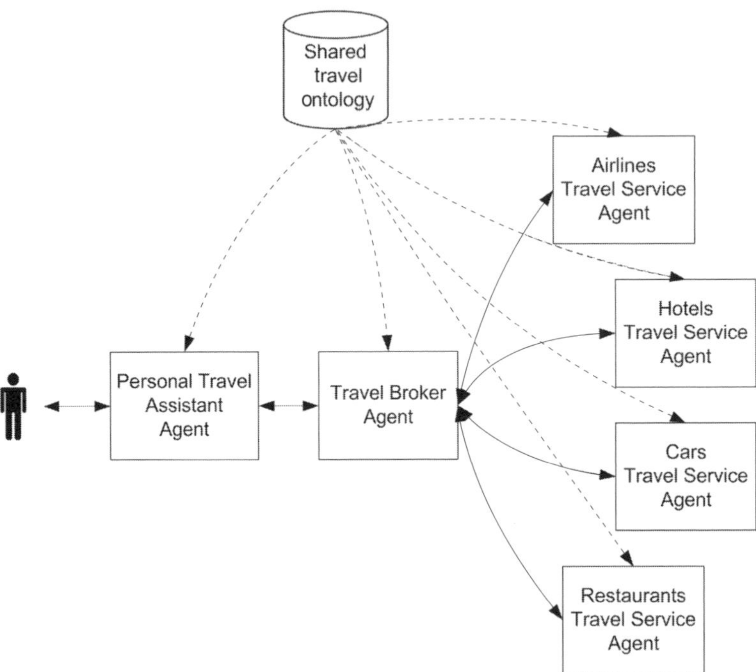

Figure 7.6 FIPA's Personal Travel Assistant agent architecture with shared ontology.

are common in standard bodies, an example of which was performed in the WAP Forum to define mobile terminal capabilities in the UAProf specification [16]. In this specification terminal attributes were defined for major terminal components that include the hardware platform, the software platform, network connectivity support, the browser, WAP features support, and finally push support. Holsapple and Joshi in [17] have reviewed additional approaches that can be taken to design an ontology, namely by adopting general principles that can then be applied to a specific ontology case, and by synthesis of a base set of ontologies.

Ontologies usually describe concepts that are referred to as classes. To provide for hierarchical descriptions, classes can be subclasses of other classes when they represent more specific concepts. Each concept contains attributes, also referred to as properties or slots of the classes. Slots can have different restrictions imposed upon them such as the slot value type, the allowed values, and the number of the values. For example, a slot can be of type string, float, Boolean, or enumerated. An enumerated type indicates that a slot can have a value from a predefined list of values only. A slot can also contain multiple values. For example, a *children* slot in a class representing a person will list all the children of this person. These children values are instances of person classes too. A set of individual instances of classes constitutes a knowledge base. Among the tools for creating ontologies, Protégé-2000 from Stanford University [18] provides the added flexibility that enables the storage of the ontology either in text files, a database, or in W3C's RDF format. Protégé-2000 supports the work of a knowledge engineer in gathering data from interviews with domain experts, and provides for iterative building of an ontology through an easily extendible class hierarchy representation.

The intended use of an ontology determines its domain and scope. The applications that will serve prespecified use cases influence the ontology's resultant classes and slots. The content of the classes and slots has to be sufficient to answer any information request. On the other hand, the ontology should not contain information that is not required by its intended users, as the integrity of extraneous information cannot be validated and therefore can only complicate future extensions of the ontology. This latter point is very important since ontologies evolve as more use cases are elaborated, and an ontology has to embody a flexible design that can grow over time. Detailed steps for developing an ontology are provided by Noy and McGuinness in [19].

Some recent research efforts have focused on automated ontology generation. For example, Missikoff *et al.* [20] have developed a software environment around the OntoLearn tool to build and assess a domain ontology for tourism enterprises. The OntoLearn tool captures domain terminology from available document repositories and Web documents. It then attaches meaning, or concepts, to the terms, as well as semantic relationships between terms, such as the *kind-of* generalization relation. For example, *conference room* is a kind-of *hotel facility*. This latter semantic interpretation is done with the help of the WordNet [6] lexical tool which provides textual definitions, sets of synonyms, and semantic relations between terms.

Criteria for ontology design and evaluation

There is no universally accepted set of ontology design rules, although different researchers have provided design criteria on how to model an ontology. For example, Gruber [21] has suggested that an ontology should exhibit clarity of the chosen terms, meaning that a suitable formalism is used to express the ontology, and it should be coherent and all inferred statements should be consistent. The ontology should be easily extensible and not depend on a particular representation language, and finally the scope of used terms should be limited to just those needed for the dealt-with application domain.

Other researchers, for example, Fernandez *et al.* have argued [22] the need for more engineering discipline. The OntoClean methodology developed by Guarino and Welty [23] is one research effort at attempting to provide more rigor to ontology specification. The OntoClean methodology applies philosophical analysis to evaluate ontologies. OntoClean specifies formal notions used to define metaproperties which characterize semantic aspects and impose taxonomic structure. These metaproperties are then used to validate an ontology specification and can expose incorrect modeling choices in a given ontology.

The state of ontology design, the determination of what constitutes a good ontology versus a bad one, is not an exact science. Readability and understandability are of course major concerns as these facilitate the adoption of the ontology terms, their semantics and relationships. A key aspect to evaluate in an ontology is to see if it can efficiently support problem resolution. A good starting point for ontology definition is to bind the domain in question by defining first the use cases for which the ontology will be applicable. For example, if the domain under consideration is location-based multimedia information delivery to a mobile user, a user's context ontology can be limited to apply to a defined set of use cases that include situations where the user is touring a city, is on a shopping outing, is traveling and needs navigation directions, or wants to locate public transportation facilities.

For these use cases, the appropriate content detail and content format, for example, audio, text, or video, are dependent on the mobile user's particular situation. The problem here is for the information service to determine the appropriate information and its format given a set of context conditions. If the terminology and relationships of a user's context ontology can help determine with a high probability of success the right content and its format, then the ontology could be considered as satisfactory. Otherwise, it may need to be extended with additional terms and relations. In the following section we review the use of W3C's Resource Description Format (RDF) and RDF Schema for resource description, and in the next chapter we present an ontology of context expressed in RDF and RDF Schema.

7.2 RDF and RDF Schema for ontology specification

Metadata is information about data that provides meaning about the content included in the data. Rather than requiring a human or computer to scan the data to determine its relevance, the metadata provides a condensed set of attributes whose values can help decide about the data's importance. This way there is less information that needs to be examined, and by using an accepted vocabulary for the attribute names it is possible to exchange this meta-information for coordinating different parties. Metadata also presents new opportunities for the generation of automated tools that scan, classify, and recommend data sources that meet prespecified search criteria. The end result is faster delivery of more precise information that is tailored to a user's request.

As information that describes data, metadata is also data that can reside in a number of places. It can either be embedded in the document it describes, or reside in a standalone document, or be included in protocol headers. These data descriptions are at times referred to as profiles, although this term is more often used when the metadata is structured in an agreed format, for example following the work of a standards group. As such, a profile can be used to describe the contents of a book citation according to the format agreed upon in the Dublin Core format [24]. Similarly, a profile can be used to describe a mobile phone's capabilities as was done by the WAP Forum's UAProf working group [16].

The Resource Description Framework (RDF) was developed by the World Wide Web Consortium (W3C) to create a format for making assertions about resources. Uniform Resource Identifiers (URIs) identify resources, and the assertions format is based on XML notation. Each resource assertion is composed of three objects: the resource itself referred to in RDF [25] as *subject*, a property name referred to as *predicate*, and a statement referred to as *object*. Object statements associate a resource's named properties with the values of these properties. In RDF there are only two types of data for property values: strings and URIs.

7.2.1 RDF applications

RDF has been applied in a few domains and is gaining acceptance as a means to exchange information about resource features between clients and servers. Examples of RDF resource descriptions include the WAP Forum's (now the Open Mobile Alliance, OMA [26]) specification of mobile terminal capabilities [16], and the W3C's specification of user privacy preferences [27].

With the proliferation of mobile terminals, content providers are required to take into consideration the capabilities of terminals so that they can adapt their content to the peculiarities of each terminal. For example, screen size, screen resolution, color support, output modalities (for example, text, graphics, video), input modalities (for example, keyboard, voice), supported user agents for markup display (for example, WML, cHTML, XHTML), and other constraints need to be considered before sending out displayable content. Some resource limitations can be conveyed via HTTP header fields, for example, the supported content types and character encodings on the client's terminal can be sent in separate HTTP headers [16].

HTTP headers can also specify quality factors that indicate relative degrees of preference, using a scale from 0 to 1. For example, in Listing 7.1 the *Accept* header indicates that the client prefers to receive WML markup (the assumed quality factor is 1), and then, in order of decreasing priority, HTML markup (with a 0.8 quality) and plain text (with a 0.5 quality). Similarly the *Accept-Charset* header specifies the preferred character encodings with Unicode first, followed by any other encoding.

```
[1]  Accept: text/wml; text/html; q=0.8, text/plain; q=0.5
[2]  Accept-Charset: unicode, *; q=0.8
```

Listing 7.1 HTTP Accept and Accept-Charset headers for media preferences specification

The WAP Forum's UAProf specification provides a much more comprehensive description of terminal capabilities when compared to the information sent in HTTP headers.

All that needs to be sent to a content server is a URL that points to the terminal capabilities Web page, usually stored on a server, so that minimal over-the-air communication is required.

The WAP Forum also included network access constraints as part of the terminal description. For example, network bandwidth and latency can affect the decision of what content to send. If the mobile user has a dual mode phone that can support both WLAN and cellular access, and happens to roam to a WLAN hotspot, the larger bandwidth can be leveraged to display video on the mobile terminal. Network access constraints are dynamic, and as such, this information is best kept on the mobile terminal, and sent upon request to a content server. For example, the *CurrentBearerService* property of the *NetworkCharacteristics* component defined in the UAProf schema [16] could specify whether the user is on a WLAN or a cellular network.

7.2.2 RDF graph models

The relationships between RDF elements (resources, property names, and property values) are represented as directed labeled graphs. Resources are the source nodes, the property names are the arcs, and property values are the nodes at the end of the arcs. Enriching a previously defined resource with new properties would require the addition of arcs with associated property values. For example, a phone's properties can be described by one source node representing the phone with one arc for each of the phone's properties.

RDF also enables the representation of structured information. This is made possible by the fact that RDF property values can be resources. These latter resources have, in turn, their own property values that form the structured information of the top-level resource. To follow on the above example, *blank nodes* are used to represent a phone's structured information resources and corresponding values are then associated with the property values of each blank node. Figure 7.7 uses the terminology of the UAProf schema to illustrate part of a phone's structured information organized in a hardware platform, a software platform, and a browser.

The graph shows that property values can have very different representations. To facilitate the correct interpretation of these strings, data types should be associated with properties. RDF Schema, defined in section 7.2.4, provides mechanisms that support type associations.

7.2.3 RDF XML syntax

RDF provides an XML syntax for describing the relationship graphs described in the previous section. Relationships are described by a self-contained XML code segment that lists the resource, its properties, and the corresponding property values. For

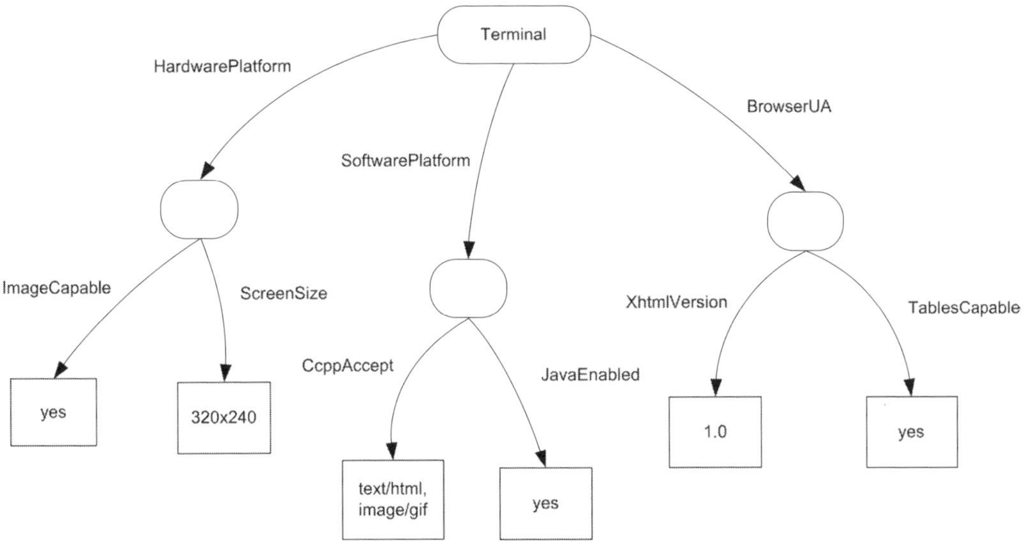

Figure 7.7 RDF graph of a phone's structured resources.

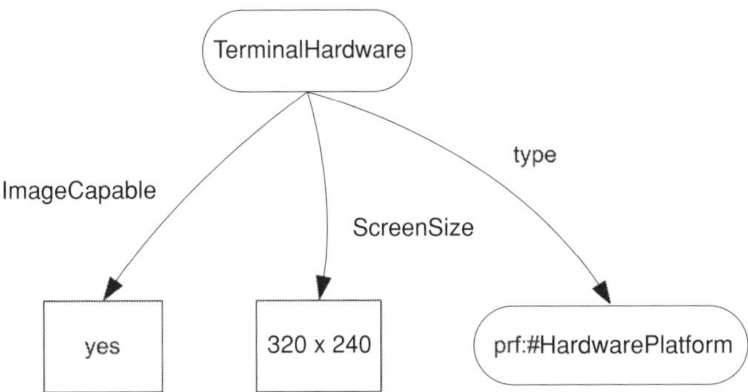

Figure 7.8 RDF sub-graph of a phone's hardware platform.

example, the Figure 7.8 shows a sub-graph of a terminal's hardware platform that specifies a phone device which can show images and has a screen size of 320 × 240 pixels (Figure 7.8).

The corresponding RDF/XML code of this graph is given in Listing 7.2.

```
[1]  <?xml version="1.0"?>
[2]  <rdf:RDF xmlns:rdf="http://www.w3.org/1999/02/22-rdf-syntax-ns#"
       xmlns:prf="http://www.wapforum.org/profiles/UAPROF/
        ccppschema-20010430#">
[3]  <rdf:Description rdf:about="http://www.phone.org/
       TerminalHardware.html">
```

```
[4]     <rdf:type resource="prf:#HardwarePlatform"/>
[5]     <prf:ImageCapable>yes</prf:ImageCapable>
[6]     <prf:ScreenSize>320x240</prf:ScreenSize>
[7]   </rdf:Description>
[8]   </rdf:RDF>
```

Listing 7.2 RDF/XML of a phone's hardware platform

The first line specifies the XML version. The *<rdf:RDF>* element (line 2) defines an RDF resource representation; the *xmlns:rdf* attribute specifies an XML namespace of W3C RDF specific terms, and the *xmlns:prf* attribute specifies the XML namespace of the WAP Forum's UAProf terms [16]. Any terms prefixed with *rdf:* are defined in the corresponding W3C URI, and any terms prefixed with *prf:* are defined in the corresponding WAP Forum URI.

The next five lines specify the actual resource and its properties. The resource is described in the *<Description>* element that contains a URI where the resource is defined (line 3). The resource properties are coded as nested elements within the resource element. This resource describes a mobile terminal's hardware, and its type is defined by the *HardwarePlatform* component of the UAProf schema (line 4). Types are elaborated upon in the following section. Property *ImageCapable* has a value *yes* (line 5), and property *ScreenSize* has value *320X240* (line 6). Finally, line 8 closes this resource code segment.

The actual description of the *ImageCapable* and the *ScreenSize* properties, as defined by the UAProf specification, is shown in Listing 7.3.

```
<rdf:Description ID="ImageCapable">
  <rdf:type rdf:resource=
    "http://www.w3.org/2000/01/rdfschema#Property"/>
  <rdfs:domain rdf:resource="#HardwarePlatform"/>
  <rdfs:comment>
    Description: Indicates whether the device supports the
    display of images. If the value is "Yes", the property
    CcppAccept may list the types of images supported.
    Type: Boolean
    Resolution: Locked
    Examples:"Yes","No"
   </rdfs:comment>
 </rdf:Description>

 <rdf:Description ID="ScreenSize">
```

```
    <rdf:type rdf:resource="http://www.w3.org/2000/01/
      rdfschema#Property"/>
    <rdfs:domain rdf:resource="#HardwarePlatform"/>
    <rdfs:comment>
     Description: The size of the device's screen in
     units of pixels, composed of the screen
     width and the screen height.
     Type: Dimension
     Resolution: Locked
     Examples:"160x160","640x480"
    </rdfs:comment>
  </rdf:Description>
```

Listing 7.3 UAProf ImageCapable and ScreenSize phone properties

In this representation, the phone properties are structured resources of a type defined by an associated *resource* URI and apply to a domain defined by another associated *resource* URI. The corresponding RDF graph represents the phone properties with nodes that have arcs for their corresponding properties. Information in the comment sections will be described more formally in future versions of the OMA's UAProf specification. If the *<comment>* elements are replaced with RDF *<value>* elements that contain the actual values of the corresponding properties, then the graph of the terminal hardware is as shown in Figure 7.9.

Property value types

Unlike programming languages, RDF has no built-in data types that specify whether a string should be interpreted as an integer, real, binary, date, etc. To facilitate the interpretation of property values, RDF provides *typed literals*. This mechanism enables the pairing of URIs with literals representing property values. For example, the string representing the date *2003-02-23* can be paired with a URI that points to the definition of date. The corresponding XML representation will include an *rdf:datatype* attribute in the property's tag that specifies a URI of the data type definition. For the above example of a mobile terminal, the corresponding XML description of a typed version date attribute is shown in Listing 7.4.

```
<?xml version="1.0"?>
<rdf:RDF xmlns:rdf="http://www.w3.org/1999/02/22-rdf-syntax-ns#"
  xmlns:phone="http://www.phone.org/terms/">
<rdf:Description rdf:about="http://www.phone.org/UserAgent">
  <phone:VersionDate
```

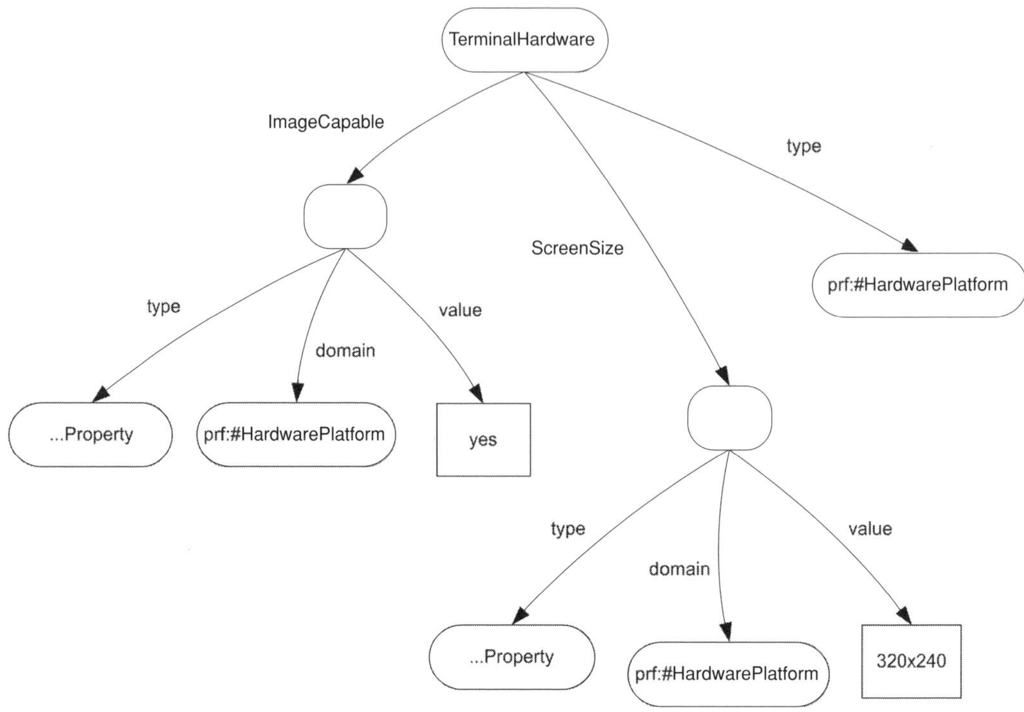

Figure 7.9 RDF sub-graph of a phone's terminal hardware with structured properties.

```
    rdf:datatype="http://www.w3.org/2001/XMLSchema#date">
     2003-02-23
    </phone:VersionDate>
  </rdf:Description>
</rdf:RDF>
```

Listing 7.4 A date property type

An RDF processor that has been programmed to understand *date* types will know how to interpret the string *2003-02-23*. As the example in Listing 7.4 shows, RDF data types can be derived from XML data types. These XML types are defined in the XML Schema specification [28]. Prior to XML schemas, DTD (Document Type Definition) documents were used to specify the allowed syntax of XML elements. DTDs specified the allowed names, and the particular hierarchical structure that the defined elements had to follow. However, DTDs have no ability to express the allowed data types of element data. This is why XML schemas were invented with included types such as *string*, *boolean*, *float*, *integer*, *date*, and *time*. Any new RDF data type is derived from these types.

Resource types

As shown in prior examples, in addition to defining property value types with *rdf:datatype*, RDF also defines *resource* types with the *rdf:type* element name. The *rdf:type* element is nested within the resource and represents a property that specifies the resource category, or class (classes will be expanded upon later). For example, the RDF XML in listing 7.5 specifies that *phone.org's UserAgent* is an instance of a *BrowserUA* class defined in the UAProf schema.

```
<?xml version=“1.0”?>
<rdf:RDF xmlns:rdf=“http://www.w3.org/1999/02/22-rdf-syntax-ns#”
     xml:base=“http://www.wapforum.org/profiles/UAPROF/
      ccppschema-20010430#”>
<rdf:Description rdf:about=“http://www.phone.org/UserAgent”>
    <rdf:type rdf:resource=“#BrowserUA”/>
</rdf:Description>
</rdf:RDF>
```

Listing 7.5 A user agent resource type

Resource groups

Resources often belong to the same group. For example, the content types that a mobile terminal can accept for display usually form an unordered list. These content types include plain text, HTML, WML, XHTML, GIF images, etc. Similarly, the preferred document languages that a mobile terminal can accept form a list too, although this is typically an ordered list. Preferred languages include English, Spanish, French, etc. To represent these groups, RDF provides predefined types for representing group containers. The RDF containers include *<Bag>* for representing an unordered group of resources, *<Sequence>* for representing an ordered group, and *<Alternative>* to represent alternative values for a property. An *<Alternative>* container is used when one out of a choice of possible values for the property can be selected.

An RDF resource is a *<Bag>* container if it includes a property with an embedded element *rdf:Bag*. The property itself is declared to have a type *Bag*. The RDF XML in Listing 7.6 specifies a *CcppAccept* property in the UAProf specification that is a *<Bag>* container for the content types that a user agent can display.

```
<?xml version=“1.0”?>
<rdf:RDF xmlns:rdf=“http://www.w3.org/1999/02/22-rdf-syntax-ns#”
  xmlns:rdfs=“http://www.w3.org/2000/01/rdf-schema#”>
<rdf:Description ID=“CcppAccept”>
  <rdf:type rdf:resource=“http://www.w3.org/2000/01/rdfschema#
   Property”/>
```

```
    <rdfs:domain rdf:resource="#SoftwarePlatform"/>
    <rdf:type rdf:resource="http://www.w3.org/2000/01/
      rdf-schema#Bag"/>
  </rdf:Description>
</rdf:RDF>
```

Listing 7.6 UAProf's CcppAccept Bag container property

The UAProf-defined *CcppAccept* property can be used in specific profiles. For example, the RDF XML in Listing 7.7 refers to UAProf's namespace *prf* and specifies a phone device that can accept different content types. The list of content types has a separate line item (*rdf:li*) for each type.

```
<?xml version="1.0"?>
<rdf:RDF xmlns:rdf="http://www.w3.org/1999/02/22-rdf-syntax-ns#"
         xmlns:prf="http://www.wapforum.org/profiles/UAPROF/
          ccppschema-20010430#">
<rdf:Description rdf:about="#CoolPhone100">
  <prf:CcppAccept>
   <rdf:Bag>
    <rdf:li rdf:resource="text/plain"/>
    <rdf:li rdf:resource="text/html"/>
    <rdf:li rdf:resource="text/wml"/>
    <rdf:li rdf:resource="text/xhtml"/>
    <rdf:li rdf:resource="image/gif"/>
   </rdf:Bag>
  </prf:CcppAccept>
 </rdf:Description>
 </rdf:RDF>
```

Listing 7.7 specific phone with a Bag container property

A *<Sequence>* container is comparable to a *<Bag>* container, except that the *rdf:Bag* element is replaced with *rdf:Seq*. In a similar way, an *<Alternative>* container uses an *rdf:Alt* element. It is up to the processing application to interpret correctly the meaning of these different containers.

7.2.4 RDF vocabularies and schemas

The collection of resources and properties that are assigned to these resources define a "vocabulary". When the type of relationships between resources and properties are defined, then we end up with what is referred to as an "RDF schema" [29]. While

RDF provides a framework for defining resource-describing terms, the domain experts define the actual terms themselves. For example, mobile manufacturers will define the terms used to describe mobile terminal capabilities. The vocabulary of specific domain knowledge that is contained in a formalized RDF schema is also referred to as an ontology.

A typical RDF document specifies what previously defined schemas are reused in the present document. As shown in prior examples, RDF schemas are usually accessible via a URI that is specified in a *namespace*. Namespaces are used to disambiguate the use of resource and property names when resources from more than one domain are manipulated. Namespaces are declared at the beginning of the document, where a namespace refers to a schema and each namespace is identified by a unique URI. The following declaration from the previous examples specifies the default RDF namespace followed by the namespace for mobile terminal capabilities and preferences defined in the WAP Forum's UAProf schema:

```
<rdf:RDF xmlns:rdf="http://www.w3.org/1999/02/22-rdf-syntax-ns#"
         xmlns:prf="http://www.wapforum.org/profiles/UAPROF/
          ccppschema-20010430#">
```

Any resource and property names prefixed by *prf:* are names reused from the UAProf schema. Using namespace prefixes ensures that names will be globally unique within the RDF document so that a schema author doesn't have to worry about potential name collisions with identical names used in other schemas. Typically, a namespace will be specified in a Web page that contains the intended meaning of the XML tags. The namespace URI is then the Web page URL.

The representation of resources from a certain domain, and their attributes, often requires an associated framework in which the resources are defined. This framework has a corresponding structure that is often hierarchical in nature with rules of inheritance and specialization. Inheritance and specialization are standard notions of object-oriented representation used in software, and are also applied in frame-based languages used in the field of artificial intelligence for defining ontologies. In an RDF schema, for example, the representation of a special purpose browser can inherit from the generic browser representation the notions of HTML version and JavaScript support. Specialized attributes of the browser could include the browser's support for particular script languages. The result of extending RDF with a framework for schema definitions is the RDF Schema vocabulary defined in the namespace *rdfs*. The following sections expand the RDF's schema definition framework.

7.2.5 RDF schema language

The facilities for defining RDF vocabularies are specified in RDF Schema (RDFS) [29], a basic type system that extends RDF with resource types and property types, and allows relationships between resources, between properties, and between properties and their associated resources, to be expressed.

RDFS classes

The basic resources that can be described in RDFS are referred to as *classes*, and specific instances of these resources can be defined as well. A class is a resource with an *rdf:type* property whose value is *rdfs:Class*. A specific instance is a resource with an *rdf:type* property whose value is the resource class name or its URI reference. A resource may be an instance of more than one class.

RDFS enables the specification of relations between classes through *subclassing*. An RDF resource can be a subclass of (that is, inherit all the properties of) a parent resource if it has a property *rdfs:subClassOf* whose value is the parent resource. Subclassing is a transitive relationship, so that chains of specialized resources can be defined. Unlike object-oriented languages such as Java where a class can have only one direct superclass, RDFS supports multiple inheritance and any class can be a subclass of more than one resource.

An example class hierarchy includes a class *MobileTerminal* that can have specialized subclasses (*Phone*, *PDA*). A multi-function mobile terminal *MultiFunctionMobile*, that is both a phone and a PDA, can be a subclass of both the *Phone* and *PDA* classes. Figure 7.10 gives an RDFS graph to illustrate this example.

The corresponding RDFS code is:

```
<?xml version="1.0"?>
<rdf:RDF
  xmlns:rdf="http://www.w3.org/1999/02/22-rdf-syntax-ns#"
  xmlns:rdfs="http://www.w3.org/2000/01/rdf-schema#">
<rdf:Description rdf:ID="MobileTerminal">
  <rdf:type rdf:resource="http://www.w3.org/2000/01/
    rdf-schema#Class"/>
</rdf:Description>
<rdf:Description rdf:ID="Phone">
  <rdf:type rdf:resource="http://www.w3.org/2000/01/
    rdf-schema#Class"/>
  <rdfs:subClassOf rdf:resource="#MobileTerminal"/>
</rdf:Description>
<rdf:Description rdf:ID="PDA">
  <rdf:type rdf:resource="http://www.w3.org/2000/01/
    rdf-schema#Class"/>
  <rdfs:subClassOf rdf:resource="#MobileTerminal"/>
</rdf:Description>
<rdf:Description rdf:ID="MultiFunctionMobile">
```

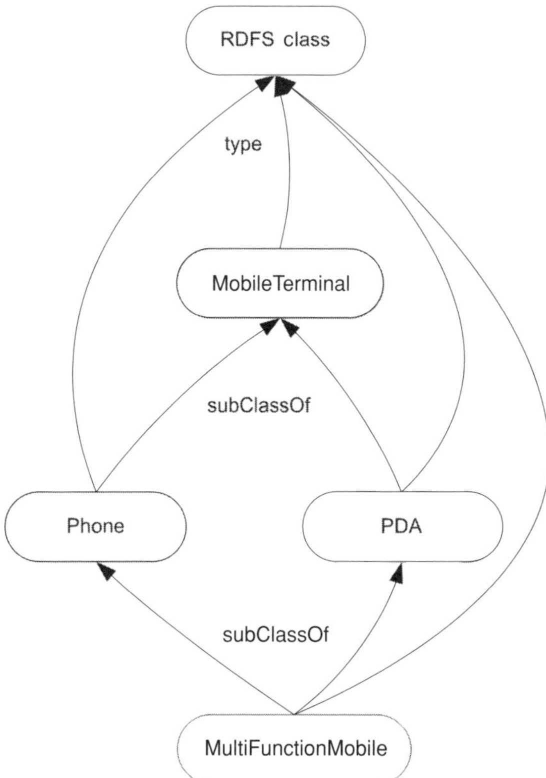

Figure 7.10 RDFS graph of class hierarchy.

```
    <rdf:type rdf:resource="http://www.w3.org/2000/01/
     rdf-schema#Class"/>
    <rdfs:subClassOf rdf:resource="#Phone"/>
    <rdfs:subClassOf rdf:resource="#PDA"/>
  </rdf:Description>
  </rdf:RDF>
```

Listing 7.8 Mobile terminals class hierarchy

Resource instances of the example mobile terminal classes are defined by an *rdf:type* property with a *resource* attribute whose value is a class name. Examples of specific instances of Phone and PDA resources are provided in Listing 7.9.

```
<?xml version="1.0"?>
<rdf:RDF
  xmlns:rdf="http://www.w3.org/1999/02/22-rdf-syntax-ns#"
  xmlns:rdfs="http://www.w3.org/2000/01/rdf-schema#">
```

```
<rdf:Description rdf:ID="Motorolai60c">
  <rdf:type rdf:resource="#Phone"/>
</rdf:Description>
<rdf:Description rdf:ID="HPJornada540">
  <rdf:type rdf:resource="#PDA"/>
</rdf:Description>
</rdf:RDF>
```

Listing 7.9 Mobile terminals class hierarchy

RDF provides an XML abbreviation syntax for describing resource instances. The *type* property is removed, and the *<Description>* element name is replaced with the class name. Our example of resource instances is now abbreviated (Listing 7.10) to includes the specification of a new namespace *ex* where the class names are defined:

```
<?xml version="1.0"?>
<rdf:RDF
  xmlns:rdf="http://www.w3.org/1999/02/22-rdf-syntax-ns#"
  xmlns:rdfs="http://www.w3.org/2000/01/rdf-schema#"
  xmlns:ex="http://www.examples.org/schema#">
<ex:Phone rdf:ID="Motorolai60c"/>
<ex:PDA rdf:ID="HPJornada540"/>
</rdf:RDF>
```

Listing 7.10 Abbreviated syntax of mobile terminals class instances

RDFS properties and property constraints

RDF Schema enables one to express properties and constraints in the relationships between resources and properties. Properties are RDF resources with an *rdf:type* property whose value is the resource *rdf:Property*. Property constraints specify the *domain* of properties and their allowed *range*. A property *domain* constrains a property to be associated with certain resources, and a property's *range* defines the allowed values that a property can have. The value of a *range* property is always a *class*. For example, in Listing 7.11 the property *ScreenSize* can have the domain *MobileTerminal*, and range of *ScreenRectangle* with possible values *SR160x160*, or *SR320x240*. These latter two values are instances of the *ScreenRectangle* class.

```
<?xml version="1.0"?>
<rdf:RDF
  xmlns:rdf="http://www.w3.org/1999/02/22-rdf-syntax-ns#"
  xmlns:rdfs="http://www.w3.org/2000/01/rdf-schema#">
<rdf:Description rdf:ID="ScreenSize">
```

```
    <rdf:type rdf:resource="http://www.w3.org/1999/02/
     22-rdf-syntax-ns#Property"/>
    <rdfs:domain rdf:resource="#MobileTerminal"/>
    <rdfs:range rdf:resource="#ScreenRectangle"/>
</rdf:Description>
</rdf:RDF>
```

Listing 7.11 ScreenSize property description

As in the case of classes, properties can be organized in a class hierarchy with the *rdfs:subPropertyOf* property. The meaning of this is that if class instances are associated with a sub-property, then they are implicitly associated with the parent property. For example, if a phone has a property *GIFcapable* that is defined in this context as a sub-property of UAProf's *ImageCapable* property, then this particular phone belongs to the category of phones that are capable of displaying images.

Component design with RDFS

RDF Schema is a property-centric specification that has many benefits. A property is an RDF resource defined independently of the classes to which it can apply. As a result, properties may be reused between classes without the need to redefine them as is done in object-oriented programming languages. This, though, requires greater care in the initial specification of properties, as future property uses may need to be considered at the outset.

Another benefit is the ability to define class instances that do not include all applicable properties. For example, a terminal's browser instance can be defined with or without a specified *XhtmlVersion* property. This is to be contrasted with object-oriented languages, such as Java, that will not allow a class instance to be created without a corresponding specification of each of its attribute values. Similarly, new properties can be added to *Terminal* device instances (see Figure 7.7) without affecting other instances of the same class that do not have these added properties. This addition of properties can be done without redefining the original resource description.

7.2.6 Beyond RDF Schema: the Web Ontology Language (OWL)

While RDFS can be used to develop basic ontologies, there are cases where the representation of relationships between objects of an ontology may require more elaboration than provided in RDFS. Further inferences can be deduced with a detailed ontology that expands the description of objects, their properties, and their relationships. In particular, the automated discovery and execution of Web services requires expressive descriptions [30]. To answer this need, the DARPA Agent Markup Language (DAML) program was started in 2000 with the goal of developing a language and tools to facilitate the concept of the Semantic Web [31]. The DAML program released an initial ontology language referred to as DAML-ONT in October 2000. Another independent effort was started to create a Web-based representation and inference layer for ontologies referred to as the Ontology Inference Layer (OIL) [32]. OIL provides more formal semantics and reasoning support which it inherits from description logics developed in knowledge-representation research.

The DAML program and the OIL effort combined to deliver the DAML+OIL ontology specification language published as a W3C note in 2001 [33]. DAML+OIL provides for more expressiveness, for example, classes can be defined to be disjoint or equivalent. Classes can also be defined by an enumeration of their instances. New semantics were defined for properties, for example, restrictions can be imposed on the cardinality of values that properties can have. Some RDFS features were also augmented, for example, a *property* could have more than one class *range*, meaning that its values are instances of multiple classes. Since then, this latter feature was integrated in the RDFS specification.

The W3C proceeded to develop a specification of the Web Ontology Language (OWL), published as a W3C recommendation in 2004 [34], [35]. This recommendation has made an interesting attempt to satisfy the needs of different communities of users and developers, and specifies three increasingly expressive sublanguages that could be used for different levels of ontology sophistication. OWL Lite extends RDFS, however, to include relatively simple constraints. For example, the cardinality of properties is limited to just the values 0 or 1, that is properties cannot have multiple values. The next levels of languages are OWL Description Logics (DL) and OWL Full. OWL DL requires strict type separation between classes, properties, and instances, while OWL Full does not have any limitations. These latter two languages use the same vocabulary and allow for more expressiveness, for example, classes can be stated to be disjoint or can be combined with Boolean relations such as union and complement relationships. There is also no restriction on the cardinality values of properties.

Ontology work within DARPA continued also with the Semantic Web Services arm of the DAML program developing an OWL-based Web Service Ontology, OWL-S (formerly referred to as DAML-S), as well as supporting tools and agent technology to enable automation of services on the Semantic Web [29], [30], [31]. OWL-S defines a core set of markup language constructs for describing the properties and capabilities of Web services. This markup of Web services descriptions is intended to facilitate the automation of tasks such as Web service discovery, invocation, composition, interoperation, and execution monitoring. Web service discovery involves the automatic location of Web services that offer specific services. Web service invocation is the automatic execution of a Web service based on markup information that describes the required input, what functions to invoke, and what formats to follow for the invocation. Interoperation and composition allows for multiple Web services to act in concert to deliver a composite service. Finally, execution monitoring tracks the status of Web service execution so that users are made aware of the progress of their requests.

To enable the above automation of Web services tasks, OWL-S defines new classes that provide service profiles with high-level service descriptions, model information on how the service works, whether composition of services needs to be enacted, and grounding information on how to programmatically access the service. For service composition, in particular, markup composition templates were defined to describe sequential, unordered, and concurrent execution of processes. In addition, precondition and effect properties can be specified for each process to encode the preconditions for performing a Web service and the effects that a service will have after it completes its execution. Also defined by OWL-S is conditional execution markup that allows the expression of conditions that should hold for a Web service to continue its execution.

Other industry efforts described in a later chapter have addressed the automation of Web service discovery and grounding, for example, UDDI specifies the structure and access to service registries, and WSDL defines an XML description of programmatic access to Web services. Few have addressed the composition of services, and this latter aspect, as described above, is elaborated in a comprehensive manner in the OWL-S specification.

REFERENCES AND FURTHER READING

[1] T. Gruber, A translation approach to portable ontology specifications. *Knowledge Acquisition*, **5**:2 (1993), 199–220.

[2] A. E. Campbell and S. C. Shapiro, Ontological mediation: An overview. *Proc. of IJCAI Workshop on Basic Ontological Issues in Knowledge Sharing*. (Menlo Park, CA: AAAI Press, 1995).

[3] M. Gruninger and J. Lee, Ontology applications and design. *Comm. of the ACM*, **45**:2 (Feb. 2002).

[4] RosettaNet, http://rosettanet.org/RosettaNet/Rooms/DisplayPages/LayoutInitial.

[5] Systematized Nomenclature of Medicine (SNOMED), http://www.snomed.org/.

[6] Princeton University, *WordNet: A Lexical Database for the English Language*, Cognitive Science Laboratory, http://www.cogsci. princeton. edu/~wn/.

[7] T. Berners-Lee *et al.*, The semantic web. *Scientific American* (May 2001).

[8] J. Heflin and J. Hendler, Searching the web with SHOE. *AAAI-2000 Workshop on AI for Web Search* (2000).

[9] A. E. Campbell *et al., Ontological Mediation: An Analysis*, Dept. of Comp. Sc., State University of New York at Buffalo. Feb. 1, 1995.

[10] G. Wiederhold and M. Genesereth, The conceptual model for mediation services. *IEEE Expert* (Sep.–Oct. 1997), 38–47.

[11] G. Wiederhold and J. Jannink, Composing diverse ontologies. *8th Working Conference on Database Semantics (DS-8)*, Rotorua, New Zealand (1999).

[12] FIPA, *FIPA Abstract Architecture Specification*, XC00001K. Nov. 2002.

[13] FIPA, *FIPA Ontology Service Specification*, XC00086D. Aug. 2001.

[14] FIPA, *FIPA Device Ontology Specification*, XC00091C. May 2002.

[15] FIPA, *FIPA Personal Travel Assistance Specification*, PC00080. Jun. 2000.

[16] WAP Forum, *WAG UAProf.* WAP-248-UAPROF-20011020-a. (http://www.openmobilealliance.org/tech/affiliates/wap/wapindex.html, Oct. 20, 2001).

[17] C. W. Holsapple and K. D. Joshi, A collaborative approach to ontology design. *Comm. of the ACM*, **45**:2 (Feb. 2002).

[18] Stanford University, *Protégé 2000*, http://protege.stanford.edu/.

[19] N. F. Noy and D. L. McGuiness, *Ontology Development 101: A Guide to Creating Your First Ontology*, Stanford Knowledge Systems Laboratory Technical Report KSL-01-05, and Stanford Medical Informatics Technical Report SMI-2001-0880. Mar. 2001.

[20] M. Missikoff *et al.*, Integrated approach to web ontology learning and engineering. *IEEE Computer* (Nov. 2002), 60–3.

[21] T. R. Gruber, Toward principles of the design of ontologies used for knowledge sharing. *Int. J. Human–Computer Studies*, **43** (1995), 907–28.

[22] M. Fernandez *et al.*, Methontology: From ontological art toward ontological engineering. *Workshop on Ontological Engineering*, Spring Symposium Series, AAAI97, Stanford. (1997).

[23] N. Guarino and C. Welty, Evaluating ontological decisions with OntoClean. *Comm. of the ACM*, **45**:2 (Feb. 2002), 61–5.

[24] Dublin Core Metadata Initiative, Citation Working Group, http://dublincore.org/groups/citation/.

[25] O. Lassila and R. Swick, *Resource Description Framework (RDF) Model and Syntax Specification*. W3C Recommendation (Feb. 22, 1999), http://www.w3.org/TR/1999/REC-rdf-syntax-19990222/.

[26] Open Mobile Alliance, http://www.openmobilealliance.org/.

[27] B. McBride *et al., An RDF Schema for P3P*. W3C Note (Jan. 25, 2002), http://www.w3.org/TR/p3p-rdfschema/.

[28] H. S. Thompson *et al., XML Schema Part I: Structures.* W3C Recommendation (May 2, 2001), http://www.w3.org/TR/xmlschema-1/.

[29] D. Brickely and R. V. Guha, *RDF Vocabulary Description Language 1.0: RDF Schema.* W3C Recommendation (Feb. 10, 2004), http://www.w3.org/TR/rdf-schema/.

[30] S. A. McIlraith *et al.*, Semantic web services. *IEEE Intelligent Systems* (Mar–Apr. 2001), 46–53.

[31] DARPA, *The DARPA Agent Markup Language (DAML)*, http://www.daml.org/index.html.

[32] Ontology Inference Layer (OIL), http://www.ontoknowledge.org/oil/oilhome.shtml.

[33] D. Connolly *et al., DAML + OIL (March 2001) Reference Description.* W3C Note (Dec. 18, 2001), http://www.w3.org/TR/daml + oil-reference.

[34] P. F. Patel-Schneider *et al., OWL Web Ontology Language Semantics and Abstract Syntax.* W3C Recommendation (Feb. 10, 2004), http://www.w3.org/TR/owl-semantics/.

[28] M. K. Smith *et al., OWL Web Ontology Language Guide*. W3C Recommendation (Feb. 10, 2004), http://www.w3.org/TR/owl-guide/.

[29] DARPA, *DAML Services*, http://www.daml.org/services/.

[30] A Ankolekar *et al.*, DAML-S: Semantic markup for web services. *Proc. of the Int. Semantic Web Working Symp. (SWWS).* (Jul. 2001).

[31] A. Ankolekar *et al.*, DAML-S: Web service description for the semantic web. *1st Int. Semantic Web Conf. (ISWC).* (Jun. 2002).

Amazon, http://www.amazon.com.

R. Fielding *et al., Hypertext Transfer Protocol – HTTP/1.1*, RFC 2616. IETF (Jun. 1999).

C. W. Holsapple and K. D. Joshi, Description and analysis of existing knowledge management frameworks. *Proc. of the Hawaiian Int. Conf. System Science*, Maui. (Jan. 1999).

H. Kim, Predicting how ontologies for the semantic web will evolve. *Comm. of the ACM*, 45:2 (Feb. 2002).

M. Marchiori *et al., Platform for Privacy Preferences (P3P) Project*. W3C Recommendation (Apr. 16, 2002), http://www.w3.org/P3P/.

8 Ontology of mobile user context

The ontology of context defined in this chapter allows one to express a user's context in a set of instances of the ontology's classes. These instances can be delivered to a wireless Web information service to adapt delivered content in much the same way as device constraints are conveyed to content providers using the W3C CC/PP framework to adapt content to a mobile device's characteristics.

We start by providing a motivation for the definition of a mobile user's context. Four major scenario categories of wireless Internet access by mobile users are selected to represent typical use cases. For each user scenario, the pertinent context elements are detailed and the user's behavior is described. In addition, for each scenario, we describe the activity of a corresponding context-aware service that delivers Web-based content. Both static and dynamic context elements are presented in the form of RDFS graphs and RDFS code.

8.1 Why an ontology of context?

Appropriate personalization of Web-based content is dependent upon the judicious collection and application of a user's context. The user's mobile terminal, profile services that store user preferences, and sensors that track external environment conditions, can all send context information to content provider servers (see Figure 8.1). Content providers, when informed about a user's particular circumstances will know how to aptly adapt their delivered content.

Mobile users access the wireless Web for different purposes that can range from network stored information retrieval to user tracking and communication. As described in the following, we have chosen to partition mobile Web use cases scenarios into four major categories. These categories are:

- City touring and entertainment.
- Shopping.
- Travel.
- User tracking and communication.

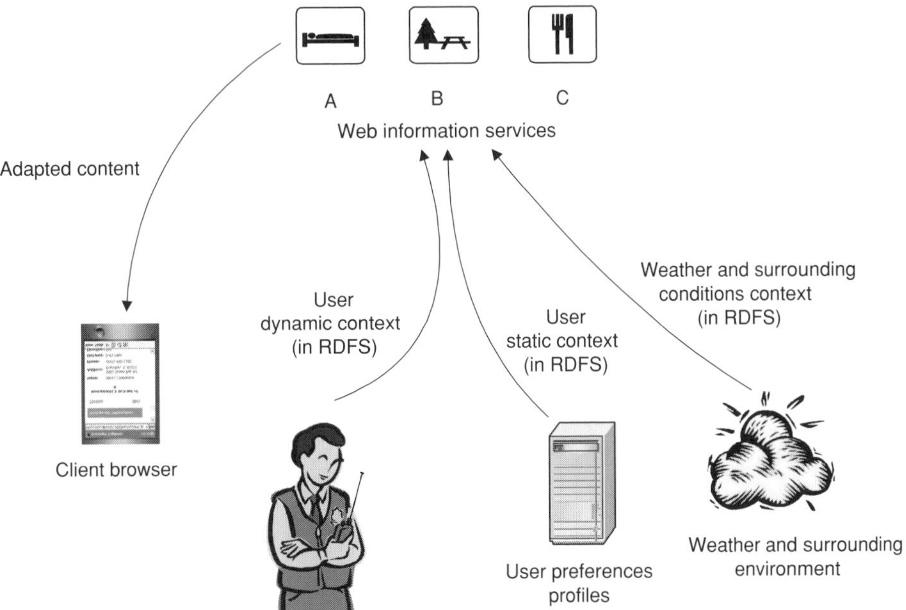

Figure 8.1 Conveying RDFS context for content adaptation.

While this grouping is not inclusive of all possible use cases, it includes major examples of usage that will drive the evolution of the wireless Internet. Although the last scenario category, user tracking and communication, does not entail access to Web-based content about city landmarks and businesses, it is included here since a user's context will affect the choice of communication mode between the user and a calling party.

How is context data used to reason about what content to deliver and when to deliver it? A mobile Web information service can use context data to "filter" the content sent to a user's mobile terminal. For example, if the user previously specified a preference for certain stores, only these will be displayed on the user's terminal when he or she issues a shopping request to find stores in his or her vicinity. Context data can also be used to express "trigger" conditions. For example, a context-aware service can detect that a user is now in the proximity of movie theaters and derive, from an appropriate database, a list of movies that match the user's preferences and will start playing in the vicinity within the next hour.

An ontology of context can be central to the operation of wireless Web services that desire to support the delivery of personalized and situation-aware content. RDFS provides a common framework for expressing this context information so that communicating parties can share a common understanding of context. Beyond filtering services, more powerful applications can be conceived that are based on inferencing engines that can deduce actions based on a set of predefined rules with associated context conditions. For inference capabilities, predicate logic operands such as *and, or, not, all, some*, and *implies* is needed. For example, a predicate statement on a user's context is:

```
<user at location L> and <time is noon> implies <user wants to be
informed about restaurants in the vicinity of L that serve lunch>
```

Inference engines could leverage RDFS-expressed context conditions in such rules to deduce actions that support user tasks. In the following, we elaborate on representative use cases, their associated context elements, and on a corresponding ontology of context expressed in RDFS.

8.2 City touring and entertainment scenarios and context

A mobile user is visiting a foreign city and wishes to be informed about available tours in his vicinity. Among the dynamic attributes of context that need to be considered by a city touring application are the user's location, date, day of the week, time of day, the prevailing weather, and user time constraints. Day of the week and time of day will determine an applicable list of tours that are offered by local tour operators. The prevailing weather will affect the choice of tours; for example, walking tours will not be advised in rainy weather. Finally, the user's time constraints will be taken into account, for example, day-long tours will not be presented to a visitor who has only three hours of free time. The user's preferences for information are considered to be static context, or slow varying context, and include topics such as architecture, and history. Visitors' cost constraints reflect the maximum amount of money they are willing to spend on a city tour.

Another example in the category of touring and entertainment is of mobile users that invoke a restaurant application to find a restaurant to their liking in their vicinity. In this case the dynamic attributes are similar to the previous use case with the exception that the current weather is usually not taken into account in the restaurant list generation process. A user-specified maximum travel distance from current location is a dynamic attribute that will help determine how extensive the list of displayed restaurants should be. Another dynamic attribute, this time of the environment, is the wait time for table availability. If the restaurant application can track wait times, the list of displayed restaurants could be ordered from shortest to longest wait. The static context here includes items such as cuisine preferences, dietary requirements, and maximum amount the user is willing to pay for a meal. Figure 8.2 shows a touring and entertainment application's activity diagram for these scenarios. Static and dynamic elements of context that are associated with these scenarios are shown as well.

The RDFS graph in Figure 8.3 depicts RDFS resources of user dynamic context in the city touring and entertainment scenarios. RDFS classes are represented in rectangles, and the labeled arcs show the relationships that hold between the classes.

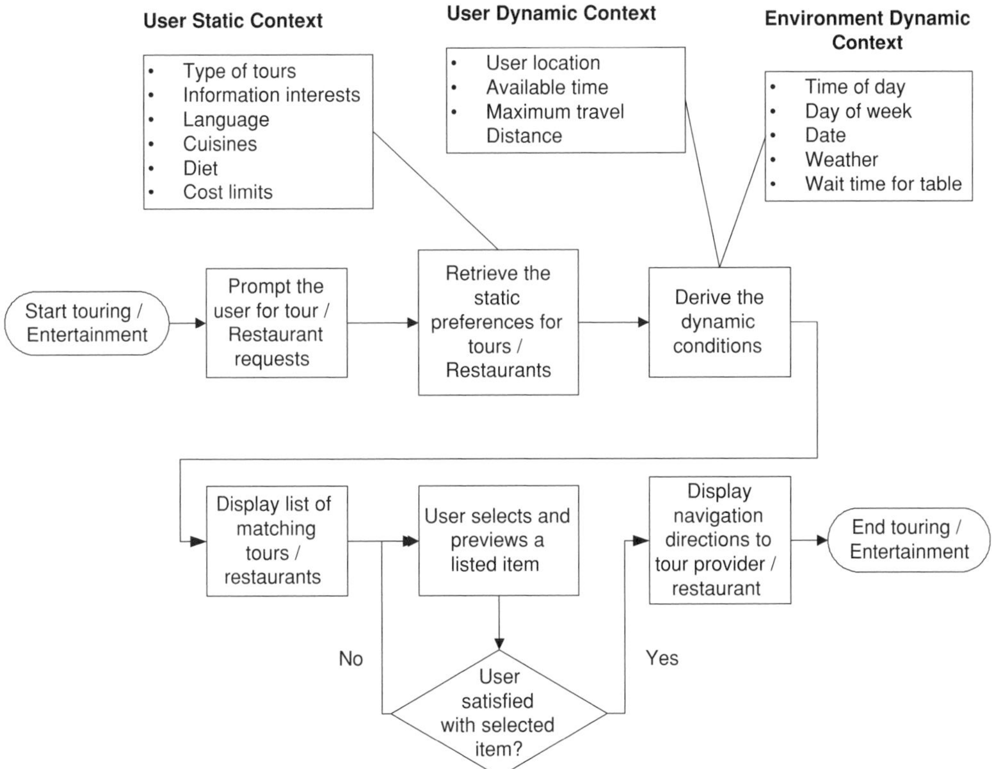

Figure 8.2 City touring / entertainment application activity diagram with example context elements.

Classes are related by RDFS properties. For example, the *User_Weather* context class is a sub-class of the context class *Dynamic*, meaning that it is a dynamic type of context. Another type of relationship is specified with *local_time*. This relationship indicates that the *local_time* property can have values from the *Environment_Time* class only. This constraint on the allowed values of a property is referred to as a "range" constraint. The *local_address* and *local_weather* relationships also indicate "range" constraints. Listing 8.1 shows the associated RDFS code of the user's dynamic context classes and properties.

```
<?xml version="1.0"?>
<rdf:RDF
  xmlns:rdf="http://www.w3.org/1999/02/22-rdf-syntax-ns#"
  xmlns:rdfs="http://www.w3.org/2000/01/rdf-schema#">

<!-- user context hierarchy -->

<rdf:Description rdf:ID="User-Context">
```

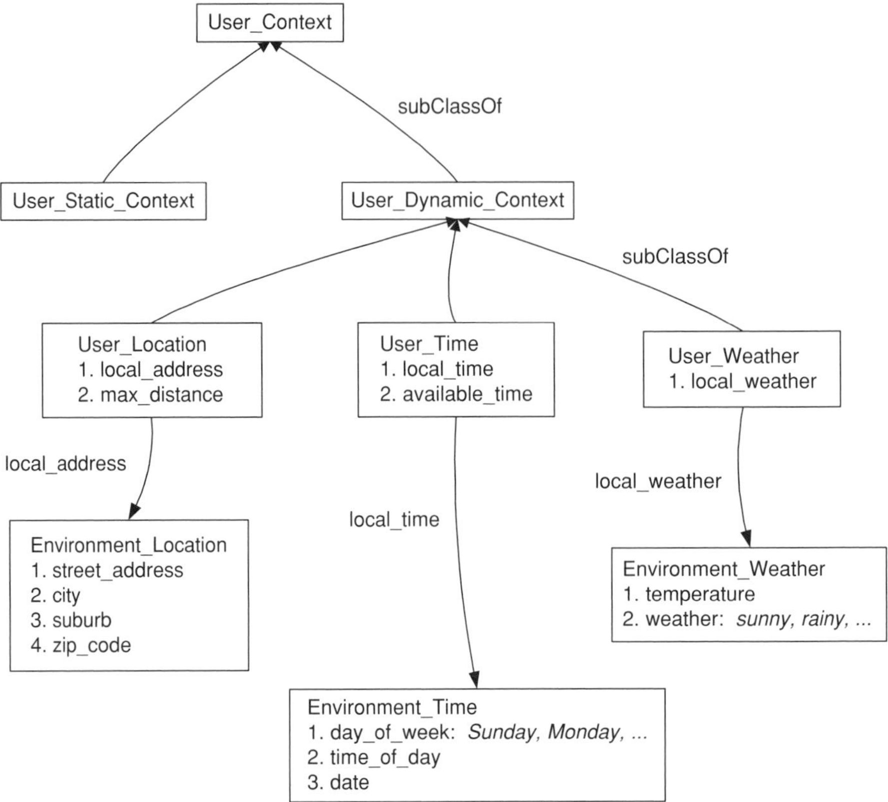

Figure 8.3 RDFS graph of user dynamic context in city touring and entertainment scenarios.

```
    <rdf:type rdf:resource="http://www.w3.org/2000/01/
     rdf-schema#Class"/>
  </rdf:Description>
  <rdf:Description rdf:ID="User-Dynamic-Context">
    <rdf:type rdf:resource="http://www.w3.org/2000/01/
     rdf-schema#Class"/>
    <rdfs:subClassOf rdf:resource="#User-Context"/>
  </rdf:Description>
  <rdf:Description rdf:ID="User-Static-Context">
    <rdf:type rdf:resource="http://www.w3.org/2000/01/
     rdf-schema#Class"/>
    <rdfs:subClassOf rdf:resource="#User-Context"/>
  </rdf:Description>

  <!-- user dynamic context classes -->

```

```
<rdf:Description rdf:ID="User-Location">
   <rdf:type rdf:resource="http://www.w3.org/2000/01/
     rdf-schema#Class"/>
   <rdfs:subClassOf rdf:resource="#User-Dynamic-Context"/>
</rdf:Description>
<rdf:Description rdf:ID="User-Time">
   <rdf:type rdf:resource="http://www.w3.org/2000/01/
     rdf-schema#Class"/>
   <rdfs:subClassOf rdf:resource="#User-Dynamic-Context"/>
</rdf:Description>
<rdf:Description rdf:ID="User-Weather">
   <rdf:type rdf:resource="http://www.w3.org/2000/01/
     rdf-schema#Class"/>
   <rdfs:subClassOf rdf:resource="#User-Dynamic-Context"/>
</rdf:Description>

<!-- properties of user dynamic context classes -->

<rdf:Description rdf:ID="local-address">
   <rdf:type rdf:resource="http://www.w3.org/1999/02/
     22-rdf-syntax-ns#Property"/>
   <rdfs:domain rdf:resource="#User-Location"/>
   <rdfs:range rdf:resource="#Environment-Location"/>
</rdf:Description>
<rdf:Description rdf:ID="max-distance">
   <rdf:type rdf:resource="http://www.w3.org/1999/
     02/22-rdf-syntax-ns#Property"/>
   <rdfs:domain rdf:resource="#User-Location"/>
   <rdfs:range rdf:resource="http://www.w3.org/2001/
     XMLSchema#decimal"/>
</rdf:Description>
<rdf:Description rdf:ID="local-time">
   <rdf:type rdf:resource="http://www.w3.org/1999/02/
     22-rdf-syntax-ns#Property"/>
   <rdfs:domain rdf:resource="#User-Time"/>
   <rdfs:range rdf:resource="#Environment-Time"/>
</rdf:Description>
<rdf:Description rdf:ID="available-time">
   <rdf:type rdf:resource="http://www.w3.org/1999/02/
     22-rdf-syntax-ns#Property"/>
   <rdfs:domain rdf:resource="#User-Time"/>
```

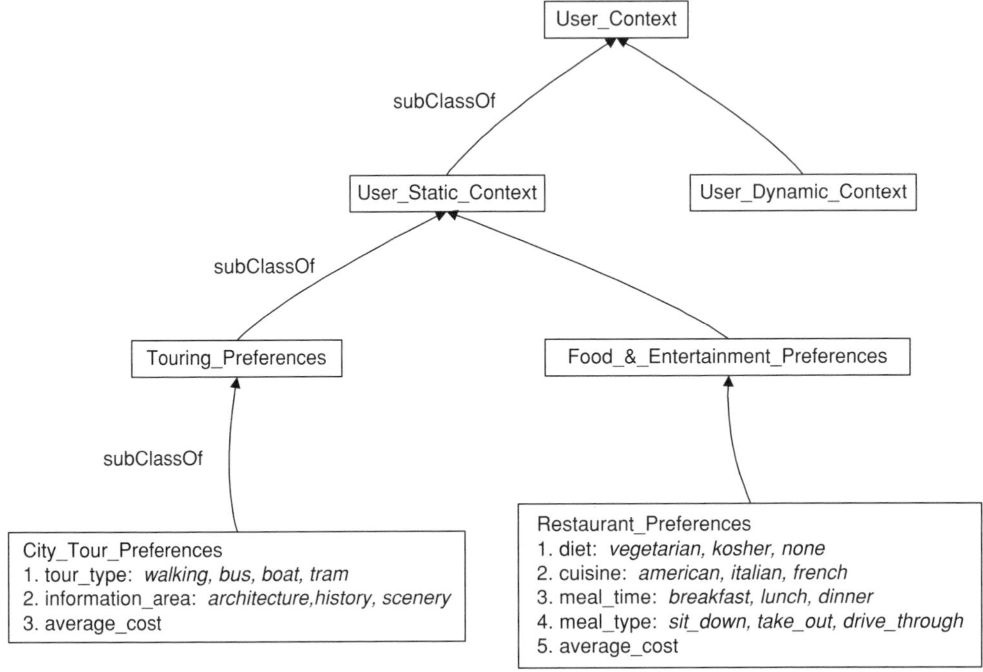

Figure 8.4 RDFS graph of user static context in city touring and entertainment scenarios.

```
    <rdfs:range rdf:resource=“http://www.w3.org/2001/
     XMLSchema#duration”/>
</rdf:Description>
<rdf:Description rdf:ID=“local-weather”>
    <rdf:type rdf:resource=“http://www.w3.org/1999/02/
     22-rdf-syntax-ns#Property”/>
    <rdfs:domain rdf:resource=“#User-Weather”/>
    <rdfs:range rdf:resource=“#Environment-Weather”/>
</rdf:Description>

</rdf:RDF>
```

Listing 8.1 RDFS code of user dynamic context in city touring and entertainment scenarios

The RDFS graph in Figure 8.4 depicts RDFS resources of user static context in the described city touring and entertainment scenarios.

Listing 8.2 shows the associated RDFS code of the user's static context classes and properties.

```
<?xml version=“1.0”?>
<rdf:RDF
 xmlns:rdf=“http://www.w3.org/1999/02/22-rdf-syntax-ns#”
 xmlns:rdfs=“http://www.w3.org/2000/01/rdf-schema#”>

<!-- user context hierarchy -->

<rdf:Description rdf:ID=“User-Context”>
   <rdf:type rdf:resource=“http://www.w3.org/2000/01/
     rdf-schema#Class”/>
 </rdf:Description>
 <rdf:Description rdf:ID=“User-Dynamic-Context”>
   <rdf:type rdf:resource=“http://www.w3.org/2000/01/
     rdf-schema#Class”/>
   <rdfs:subClassOf rdf:resource=“#User-Context”/>
 </rdf:Description>
 <rdf:Description rdf:ID=“User-Static-Context”>
   <rdf:type rdf:resource=“http://www.w3.org/2000/01/
     rdf-schema#Class”/>
   <rdfs:subClassOf rdf:resource=“# User-Context”/>
 </rdf:Description>

 <!-- user static context classes -->

 <rdf:Description rdf:ID=“Touring-Preferences”>
   <rdf:type rdf:resource=“http://www.w3.org/2000/01/
     rdf-schema#Class”/>
   <rdfs:subClassOf rdf:resource=“#User-Static-Context”/>
 </rdf:Description>
 <rdf:Description rdf:ID=“Food-Entertainment-Preferences”>
   <rdf:type rdf:resource=“http://www.w3.org/2000/01/
     rdf-schema#Class”/>
   <rdfs:subClassOf rdf:resource=“#User-Static-Context”/>
 </rdf:Description>
 <rdf:Description rdf:ID=“City-Tour-Preferences”>
   <rdf:type rdf:resource=“http://www.w3.org/2000/01/
     rdf-schema#Class”/>
   <rdfs:subClassOf rdf:resource=“#Touring-Preferences”/>
 </rdf:Description>
 <rdf:Description rdf:ID=“Restaurant-Preferences”>
```

```
  <rdf:type rdf:resource=“http://www.w3.org/2000/01/
   rdf-schema#Class”/>
  <rdfs:subClassOf rdf:resource=
   “#Food-Entertainment-Preferences”/>
</rdf:Description>

<!-- properties of user static context classes -->

<rdf:Description rdf:ID=“tour-type”>
  <rdf:type rdf:resource=“http://www.w3.org/1999/02/
   22-rdf-syntax-ns#Property“/>
  <rdf:type rdf:resource=“http://www.w3.org/2000/01/
   rdf-schema#Seq”/>
  <rdfs:domain rdf:resource=“#City-Tour-Preferences”/>
  <rdfs:range rdf:resource=“#Tour-Type-Value”/>
</rdf:Description>
<rdf:Description rdf:ID=“information-area”>
  <rdf:type rdf:resource=“http://www.w3.org/1999/02/
   22-rdf-syntax-ns#Property”/>
  <rdf:type rdf:resource=“http://www.w3.org/2000/01/
   rdf-schema#Seq”/>
  <rdfs:range rdf:resource=“#Information-Area-Value”/>
</rdf:Description>
<rdf:Description rdf:ID=“average-cost”>
  <rdf:type rdf:resource=“http://www.w3.org/1999/02/
   22-rdf-syntax-ns#Property”/>
  <rdfs:range rdf:resource=“http://www.w3.org/2001/
   XMLSchema#decimal”/>
</rdf:Description>
<rdf:Description rdf:ID=“diet”>
  <rdf:type rdf:resource=“http://www.w3.org/1999/02/
   22-rdf-syntax-ns#Property”/>
  <rdf:type rdf:resource=“http://www.w3.org/2000/01/
   rdf-schema#Alt”/>
  <rdfs:domain rdf:resource=“#Restaurant-Preferences”/>
  <rdfs:range rdf:resource=“#Diet-Value”/>
</rdf:Description>
<rdf:Description rdf:ID=“cuisine”>
  <rdf:type rdf:resource=“http://www.w3.org/1999/02/
   22-rdf-syntax-ns#Property”/>
```

```
    <rdf:type rdf:resource="http://www.w3.org/2000/01/
      rdf-schema#Seq"/>
    <rdfs:domain rdf:resource="#Restaurant_Preferences"/>
    <rdfs:range rdf:resource="#Cuisine_Value"/>
  </rdf:Description>
  <rdf:Description rdf:ID="meal_time">
    <rdf:type rdf:resource="http://www.w3.org/1999/02/
      22-rdf-syntax-ns#Property"/>
    <rdf:type rdf:resource="http://www.w3.org/2000/01/
      rdf-schema#Alt"/>
    <rdfs:domain rdf:resource="#Restaurant_Preferences"/>
  <rdfs:range rdf:resource="#Meal_Time_Value"/>
  </rdf:Description>
  <rdf:Description rdf:ID="meal_type">
    <rdf:type rdf:resource="http://www.w3.org/1999/02/
      22-rdf-syntax-ns#Property"/>
    <rdf:type rdf:resource="http://www.w3.org/2000/01/
      rdf-schema#Seq"/>
    <rdfs:domain rdf:resource="#Restaurant_Preferences"/>
    <rdfs:range rdf:resource="#Meal_Type_Value"/>
  </rdf:Description>

  </rdf:RDF>
```

Listing 8.2 RDFS code of user static context in city touring and entertainment scenarios

Some RDFS properties in this code do not have associated *domain* attributes so as not to confine them to specific resource types. As RDFS uses a property centric design approach, the same properties could then be reused in other resource types to those where they appear here. For example, the *average_cost* property is used in more than one resource type.

User preferences can be specified with RDFS properties of type *Sequence*, meaning that they can contain a prioritized list of preference values. For example, a specific resource with the *cuisine* property could list the value *american* followed by *italian*, meaning that the user prefers first American cuisine, and Italian cuisine next. In other cases, the user selects only one of a possible list of values. For example, the *diet* property is of type *Alternative* and represents a user's selected diet such as *vegetarian*. The classes and associated instances defined in Listing 8.3 show the possible values for those properties that have specified ranges.

```
<?xml version="1.0"?>
<rdf:RDF
  xmlns:rdf="http://www.w3.org/1999/02/22-rdf-syntax-ns#"
  xmlns:rdfs="http://www.w3.org/2000/01/rdf-schema#">
<!-- defined property values -->
<rdfs:Class rdf:ID="Tour-Type-Value"/>
<rdf:Description rdf:ID="walking">
    <rdf:type rdf:resource="#Tour-Type-Value"/>
</rdf:Description>
<rdf:Description rdf:ID="bus">
    <rdf:type rdf:resource="#Tour-Type-Value"/>
</rdf:Description>
<rdf:Description rdf:ID="boat">
    <rdf:type rdf:resource="#Tour-Type-Value"/>
</rdf:Description>
<rdf:Description rdf:ID="tram">
    <rdf:type rdf:resource="#Tour-Type-Value"/>
</rdf:Description>
<rdfs:Class rdf:ID="Information-Area-Value"/>
<rdf:Description rdf:ID="architecture">
    <rdf:type rdf:resource="#Information-Area-Value"/>
</rdf:Description>
<rdf:Description rdf:ID="history">
    <rdf:type rdf:resource="#Information-Area-Value"/>
</rdf:Description>
<rdf:Description rdf:ID="scenery">
    <rdf:type rdf:resource="#Information-Area-Value"/>
</rdf:Description>
<rdfs:Class rdf:ID="Diet-Value"/>
<rdf:Description rdf:ID="vegetarian">
    <rdf:type rdf:resource="#Diet-Value"/>
</rdf:Description>
<rdf:Description rdf:ID="kosher">
    <rdf:type rdf:resource="#Diet-Value"/>
```

```
</rdf:Description>
<rdf:Description rdf:ID="none">
  <rdf:type rdf:resource="#Diet-Value"/>
</rdf:Description>

<rdfs:Class rdf:ID="Cuisine-Value"/>

<rdf:Description rdf:ID="american">
  <rdf:type rdf:resource="#Cuisine-Value"/>
</rdf:Description>
<rdf:Description rdf:ID="italian">
  <rdf:type rdf:resource="#Cuisine-Value"/>
</rdf:Description>
<rdf:Description rdf:ID="french">
  <rdf:type rdf:resource="#Cuisine-Value"/>
</rdf:Description>
<rdf:Description rdf:ID="chinese">
  <rdf:type rdf:resource="#Cuisine-Value"/>
</rdf:Description>

<rdfs:Class rdf:ID="Meal-Time-Value"/>

<rdf:Description rdf:ID="breakfast">
  <rdf:type rdf:resource="#Meal-Time-Value"/>
</rdf:Description>
<rdf:Description rdf:ID="lunch">
  <rdf:type rdf:resource="#Meal-Time-Value"/>
</rdf:Description>
<rdf:Description rdf:ID="dinner">
  <rdf:type rdf:resource="#Meal-Time-Value"/>
</rdf:Description>

<rdfs:Class rdf:ID="Meal-Type-Value"/>

<rdf:Description rdf:ID="sit-down">
  <rdf:type rdf:resource="#Meal-Type-Value"/>
</rdf:Description>
<rdf:Description rdf:ID="take-out">
  <rdf:type rdf:resource="#Meal-Type-Value"/>
</rdf:Description>
<rdf:Description rdf:ID="drive-through">
```

```
    <rdf:type rdf:resource="#Meal-Type-Value"/>
</rdf:Description>

</rdf:RDF>
```

Listing 8.3 RDFS code of property values in city touring and entertainment scenarios

8.3 Shopping scenarios and context

A mobile user that plans to purchase a list of items will provide this list to a shopping application. The list of purchase items is part of the user's dynamic context. Other dynamic context elements include the maximum distance the user is willing to travel to shop, as well as the time of day and day of week used for checking store opening hours. In addition, users have preferences for certain stores, for a category of stores, for the store's location set-up, that is, whether in a mall or stand-alone, and these static preferences can help determine the list of nearby stores that will be displayed on the user's terminal.

Once in a store, if a user has previously defined product review preferences, these could help select what product information to display. For example, a location detection system could detect the user's presence in the vicinity of store exhibits that contain the items he or she may want to purchase. Relevant product reviews, or how-to-use information, could then be sent to the user's terminal. User preferences could indicate, for example, the review sources, that is, the specific sites from which the user wishes to receive product reviews.

Users can indicate in their respective profiles whether they are willing to receive shopping advertisements. In the affirmative, users can indicate the shopping categories for which they would want advertisements, for example, notifications of music album releases, and whether they are willing to receive ecoupons. Given this information, users' dynamic context can help determine the applicable sale advertisements and ecoupons that are sent to their devices. For example, the proximity of a user to a store that carries items they wish to purchase could be a trigger for sending targeted advertisements and other purchase incentives. Figure 8.5 shows a shopping and advertisement application's activity diagram for the above scenarios and the associated static and dynamic elements of context.

The RDFS graph in Figure 8.6 depicts RDFS resources of user dynamic context in the described shopping scenarios.

The associated RDFS code of the user's dynamic context classes and properties is shown is Listing 8.4.

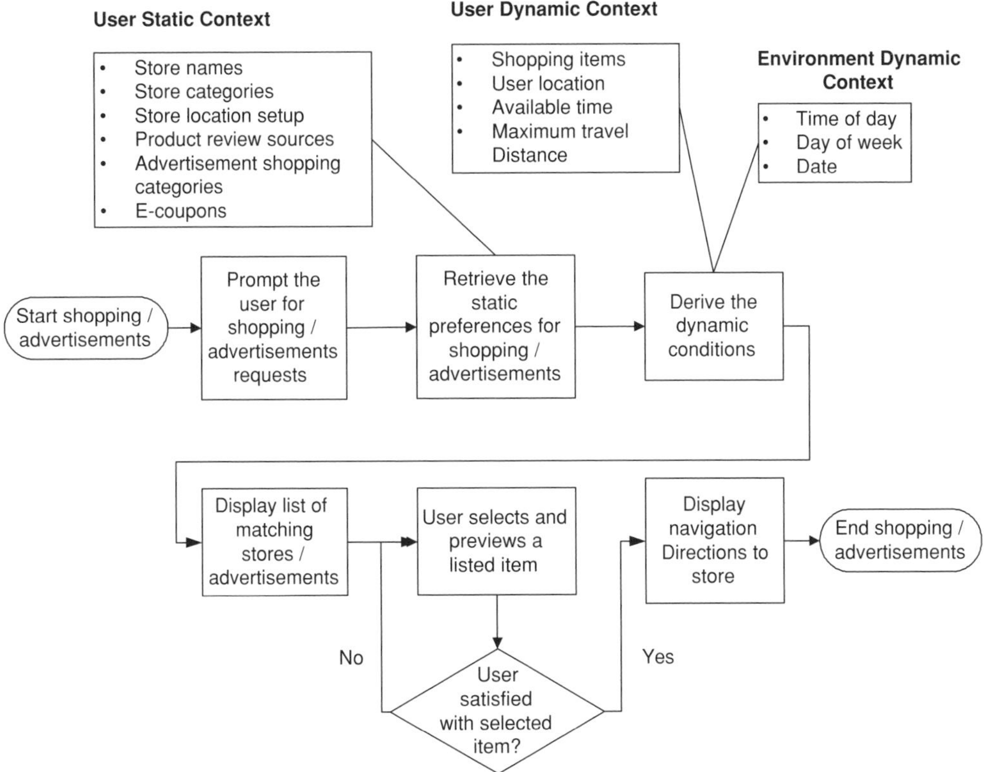

Figure 8.5 Shopping and advertisement application activity diagram with example context elements.

```
<?xml version="1.0"?>
<rdf:RDF
  xmlns:rdf="http://www.w3.org/1999/02/22-rdf-syntax-ns#"
  xmlns:rdfs="http://www.w3.org/2000/01/rdf-schema#">

<!-- user context hierarchy -->

<rdf:Description rdf:ID="User-Context">
  <rdf:type rdf:resource="http://www.w3.org/2000/01/
    rdf-schema#Class"/>
</rdf:Description>
<rdf:Description rdf:ID="User-Dynamic-Context">
  <rdf:type rdf:resource="http://www.w3.org/2000/01/
    rdf-schema#Class"/>
  <rdfs:subClassOf rdf:resource="#User-Context"/>
```

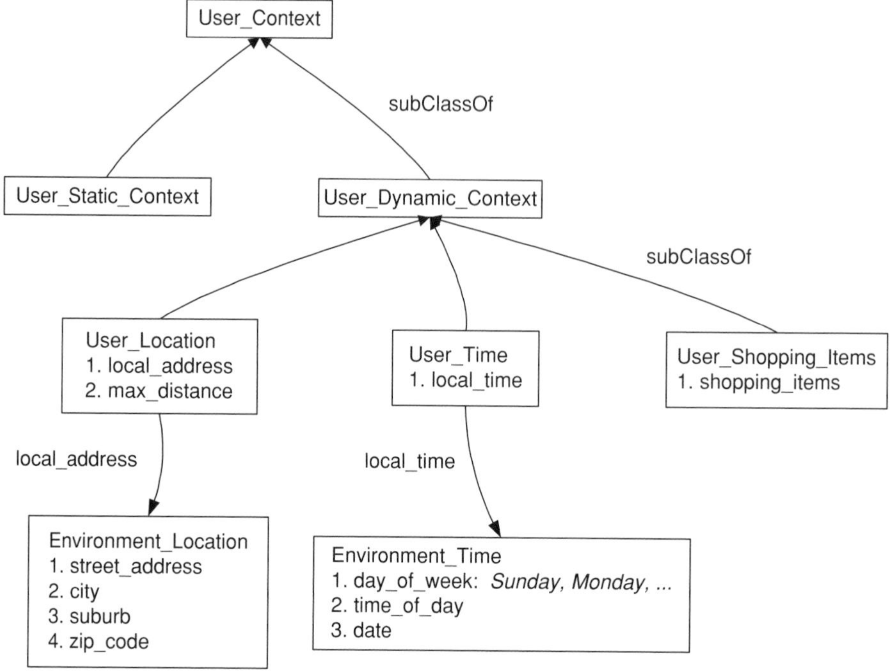

Figure 8.6 RDFS graph of user dynamic context in shopping scenarios.

```
</rdf:Description>
<rdf:Description rdf:ID="User-Static-Context">
   <rdf:type rdf:resource="http://www.w3.org/2000/01/
    rdf-schema#Class"/>
   <rdfs:subClassOf rdf:resource="#User-Context"/>
</rdf:Description>

<!-- user dynamic context classes -->

<rdf:Description rdf:ID="User-Location">
   <rdf:type rdf:resource="http://www.w3.org/2000/01/
    rdf-schema#Class"/>
   <rdfs:subClassOf rdf:resource="#User-Dynamic-Context"/>
</rdf:Description>
<rdf:Description rdf:ID="User-Time">
   <rdf:type rdf:resource="http://www.w3.org/2000/01/
    rdf-schema#Class"/>
   <rdfs:subClassOf rdf:resource="#User-Dynamic-Context"/>
</rdf:Description>
```

```
<rdf:Description rdf:ID="User-Shopping-Items">
   <rdf:type rdf:resource="http://www.w3.org/2000/01/
    rdf-schema#Class"/>
   <rdfs:subClassOf rdf:resource="#User-Dynamic-Context"/>
</rdf:Description>

<!-- properties of user dynamic context classes -->

<rdf:Description rdf:ID="local-address">
   <rdf:type rdf:resource="http://www.w3.org/1999/02/
    22-rdf-syntax-ns#Property"/>
   <rdfs:domain rdf:resource="#User-Location"/>
   <rdfs:range rdf:resource="#Environment-Location"/>
</rdf:Description>
<rdf:Description rdf:ID="max-distance">
   <rdf:type rdf:resource="http://www.w3.org/1999/02/
    22-rdf-syntax-ns#Property"/>
   <rdfs:domain rdf:resource="#User-Location"/>
   <rdfs:range rdf:resource="http://www.w3.org/2001/
    XMLSchema#decimal"/>
</rdf:Description>
<rdf:Description rdf:ID="local-time">
   <rdf:type rdf:resource="http://www.w3.org/1999/02/
    22-rdf-syntax-ns#Property"/>
   <rdfs:domain rdf:resource="#User-Time"/>
   <rdfs:range rdf:resource="#Environment-Time"/>
</rdf:Description>
<rdf:Description rdf:ID="shopping-items">
   <rdf:type rdf:resource="http://www.w3.org/1999/02/
    22-rdf-syntax-ns#Property"/>
   <rdf:type rdf:resource="http://www.w3.org/2000/01/
    rdf-schema#Bag"/>
   <rdfs:domain rdf:resource="#User-Shopping-Items"/>
   <rdfs:range rdf:resource="http://www.w3.org/2001/
    XMLSchema#string"/>
</rdf:Description>

</rdf:RDF>
```

Listing 8.4 RDFS code of user dynamic context in shopping scenarios

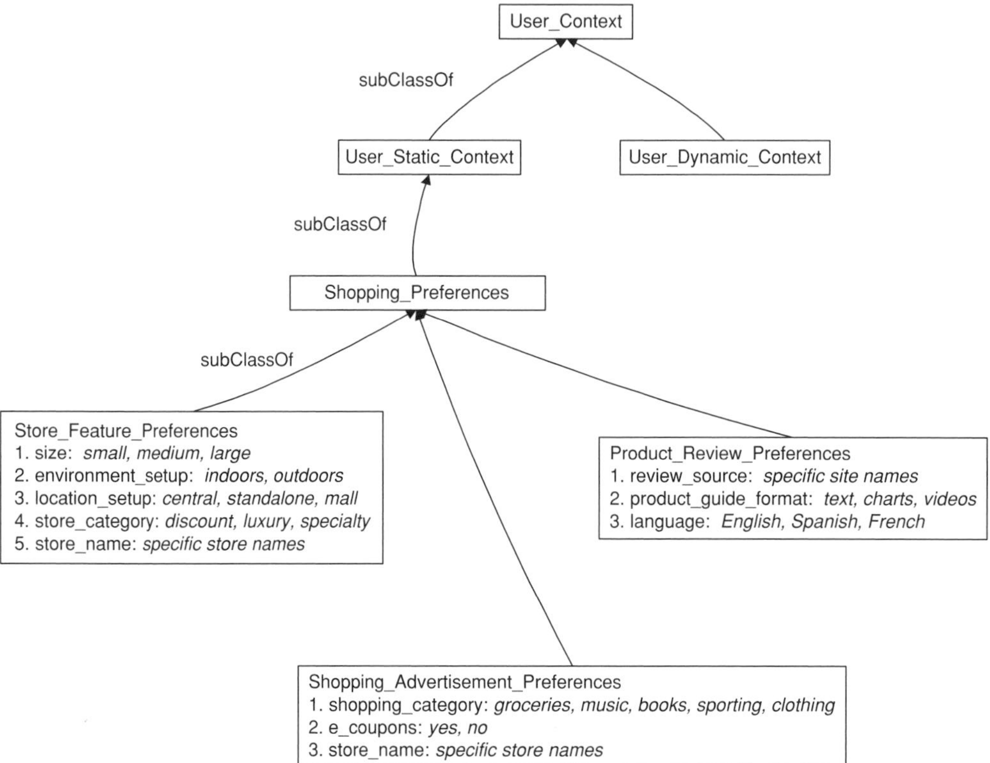

Figure 8.7 RDFS graph of user static context in shopping scenarios.

The RDFS graph in Figure 8.7 depicts RDFS resources of user static context in the described shopping scenarios.

Listing 8.5 shows the associated RDFS code of the user's static context classes and properties.

```
<?xml version="1.0"?>
<rdf:RDF
  xmlns:rdf="http://www.w3.org/1999/02/22-rdf-syntax-ns#"
  xmlns:rdfs="http://www.w3.org/2000/01/rdf-schema#">

<!-- user context hierarchy -->

<rdf:Description rdf:ID="User-Context">
  <rdf:type rdf:resource="http://www.w3.org/2000/01/
    rdf-schema#Class"/>
</rdf:Description>
<rdf:Description rdf:ID="User-Dynamic-Context">
```

```
    <rdf:type rdf:resource="http://www.w3.org/2000/01/
     rdf-schema#Class"/>
    <rdfs:subClassOf rdf:resource="#User-Context"/>
</rdf:Description>
<rdf:Description rdf:ID="User-Static-Context">
    <rdf:type rdf:resource="http://www.w3.org/2000/01/
     rdf-schema#Class"/>
    <rdfs:subClassOf rdf:resource="#User-Context"/>
</rdf:Description>

<!-- user static context classes -->

<rdf:Description rdf:ID="Shopping-Preferences">
    <rdf:type rdf:resource="http://www.w3.org/2000/01/
     rdf-schema#Class"/>
    <rdfs:subClassOf rdf:resource="#User-Static-Context"/>
</rdf:Description>
<rdf:Description rdf:ID="Store-Feature-Preferences">
    <rdf:type rdf:resource="http://www.w3.org/2000/01/
     rdf-schema#Class"/>
    <rdfs:subClassOf rdf:resource="#Shopping-Preferences"/>
</rdf:Description>
<rdf:Description rdf:ID="Product-Review-Preferences">
    <rdf:type rdf:resource="http://www.w3.org/2000/01/
     rdf-schema#Class"/>
    <rdfs:subClassOf rdf:resource="#Shopping-Preferences"/>
</rdf:Description>
<rdf:Description rdf:ID="Shopping-Advertisement-Preferences">
    <rdf:type rdf:resource="http://www.w3.org/2000/01/
     rdf-schema#Class"/>
    <rdfs:subClassOf rdf:resource="#Shopping-Preferences"/>
</rdf:Description>

<!-- properties of user static context classes -->

<rdf:Description rdf:ID="size">
    <rdf:type rdf:resource="http://www.w3.org/1999/02/
     22-rdf-syntax-ns#Property"/>
    <rdf:type rdf:resource="http://www.w3.org/2000/01/
     rdf-schema#Seq"/>
    <rdfs:range rdf:resource="#Size-Value"/>
```

```
</rdf:Description>
<rdf:Description rdf:ID=“environment-setup”>
   <rdf:type rdf:resource=“http://www.w3.org/1999/02/
    22-rdf-syntax-ns#Property”/>
   <rdf:type rdf:resource=“http://www.w3.org/2000/01/
    rdf-schema#Seq”/>
   <rdfs:range rdf:resource=“#Environment-Setup-Value”/>
</rdf:Description>
<rdf:Description rdf:ID=“location-setup”>
   <rdf:type rdf:resource=“http://www.w3.org/1999/02/
    22-rdf-syntax-ns#Property”/>
   <rdf:type rdf:resource=“http://www.w3.org/2000/01/
    rdf-schema#Seq”/>
   <rdfs:range rdf:resource=“#Location-Setup-Value”/>
</rdf:Description>
<rdf:Description rdf:ID=“store-category”>
   <rdf:type rdf:resource=“http://www.w3.org/1999/02/
    22-rdf-syntax-ns#Property”/>
   <rdf:type rdf:resource=“http://www.w3.org/2000/01/
    rdf-schema#Seq”/>
   <rdfs:domain rdf:resource=“#Store-Feature-Preferences”/>
   <rdfs:range rdf:resource=“#Store-Category-Value”/>
</rdf:Description>
<rdf:Description rdf:ID=“store-name”>
   <rdf:type rdf:resource=“http://www.w3.org/1999/02/
    22-rdf-syntax-ns#Property”/>
   <rdf:type rdf:resource=“http://www.w3.org/2000/01/
    rdf-schema#Seq”/>
   <rdfs:domain rdf:resource=“#Store-Feature-Preferences”/>
   <rdfs:range rdf:resource=“http://www.w3.org/2001/
    XMLSchema#string”/>
</rdf:Description>
<rdf:Description rdf:ID=“review-source”>
   <rdf:type rdf:resource=“http://www.w3.org/1999/02/
    22-rdf-syntax-ns#Property”/>
   <rdf:type rdf:resource=“http://www.w3.org/2000/01/
    rdf-schema#Seq”/>
   <rdfs:domain rdf:resource=“#Product-Review-Preferences”/>
   <rdfs:range rdf:resource=“http://www.w3.org/2001/
    XMLSchema#string”/>
</rdf:Description>
```

```
<rdf:Description rdf:ID="product-guide-format">
   <rdf:type rdf:resource="http://www.w3.org/1999/02/
    22-rdf-syntax-ns#Property"/>
   <rdf:type rdf:resource="http://www.w3.org/2000/01/
    rdf-schema#Seq"/>
   <rdfs:domain rdf:resource="#Product-Review-Preferences"/>
   <rdfs:range rdf:resource="#Product-Guide-Format-Value"/>
</rdf:Description>
<rdf:Description rdf:ID="language">
   <rdf:type rdf:resource="http://www.w3.org/1999/02/
    22-rdf-syntax-ns#Property"/>
   <rdf:type rdf:resource="http://www.w3.org/2000/01/
    rdf-schema#Alt"/>
   <rdfs:range rdf:resource="#Language-Value"/>
</rdf:Description>
<rdf:Description rdf:ID="shopping-category">
   <rdf:type rdf:resource="http://www.w3.org/1999/02/
    22-rdf-syntax-ns#Property"/>
   <rdf:type rdf:resource="http://www.w3.org/2000/01/
    rdf-schema#Seq"/>
   <rdfs:domain rdf:resource="#Shopping-Advertisement-
    Preferences"/>
   <rdfs:range rdf:resource="#Shopping-Category-Value"/>
</rdf:Description>
<rdf:Description rdf:ID="e-coupons">
   <rdf:type rdf:resource="http://www.w3.org/1999/02/
    22-rdf-syntax-ns#Property"/>
   <rdf:type rdf:resource="http://www.w3.org/2000/01/
    rdf-schema#Alt"/>
   <rdfs:domain rdf:resource="#Shopping-Advertisement-
    Preferences"/>
   <rdfs:range rdf:resource="#E-Coupons-Value"/>
</rdf:Description>

</rdf:RDF>
```

Listing 8.5 RDFS code of user static context in shopping scenarios

Possible values for those properties with specified ranges are listed in the RDFS class hierarchy graph in Figure 8.7. In the RDFS code in Listing 8.5, only some properties have an associated *domain*. For example, the *location_setup* property could be applied to other resources besides stores, and therefore does not include a *domain* attribute. However, the *store_category* property is specific to the description of stores and does, accordingly, include a *domain* that limits the property's use to store instances. While there are no prescriptive instructions on the use of *domain* constraint information, RDFS parsers could, for example, use them to check the validity of resource definitions.

8.4 Travel scenarios and context

When travelers wish to reach a certain destination by car, the choice of route can be influenced by their static preferences for type of route, for example, highways, city roads, scenic routes. Dynamic context of roads' traffic conditions, as tracked by road sensors, can help select roads that are not congested. Alternatively, environment context such as the current day of week and time of day can determine the expected traffic conditions, and therefore help a travel application recommend roads that are less busy. A specific route could be determined by other constraints such as the travel's purpose. For example, if the user is on the way to a store, then the travel application could recommend a less direct route which includes other stores that carry similar items to those the buyer intends to purchase.

A mobile user that requests travel directions can be the driver, a passenger sitting next to the driver, a passenger on public transportation such as a bus, or a pedestrian. Depending on the user's particular circumstances, travel directions could be sent in different formats. A driver may prefer to receive audio information that gives turn-by-turn instructions, whereas a passenger could view maps and associated visuals that provide additional information on the surrounding environment. Upon nearing a destination, a user may want to find a parking lot. Users may have specific preferences for parking their car, for example, some may prefer indoor parking to outdoor parking. These preferences are leveraged by a travel application that generates an appropriate parking lot list sent to driving mobile users.

If the mobile user intends to travel by public transportation, then the user's preferences for type of transportation are taken into account. Possible transportation choices could be bus, subway, train, cab. The user may also specify cost limits. The user's particular circumstances, for example, available time, may influence which transportation means is selected. The schedule of a specific transportation will also affect its suitability, as the schedule will determine the time of arrival at a specified destination. Figure 8.8 shows a travel application's activity diagram for these scenarios and the associated static and dynamic elements of context.

The RDFS graph in Figure 8.9 depicts RDFS resources of user dynamic context in the described travel scenarios.

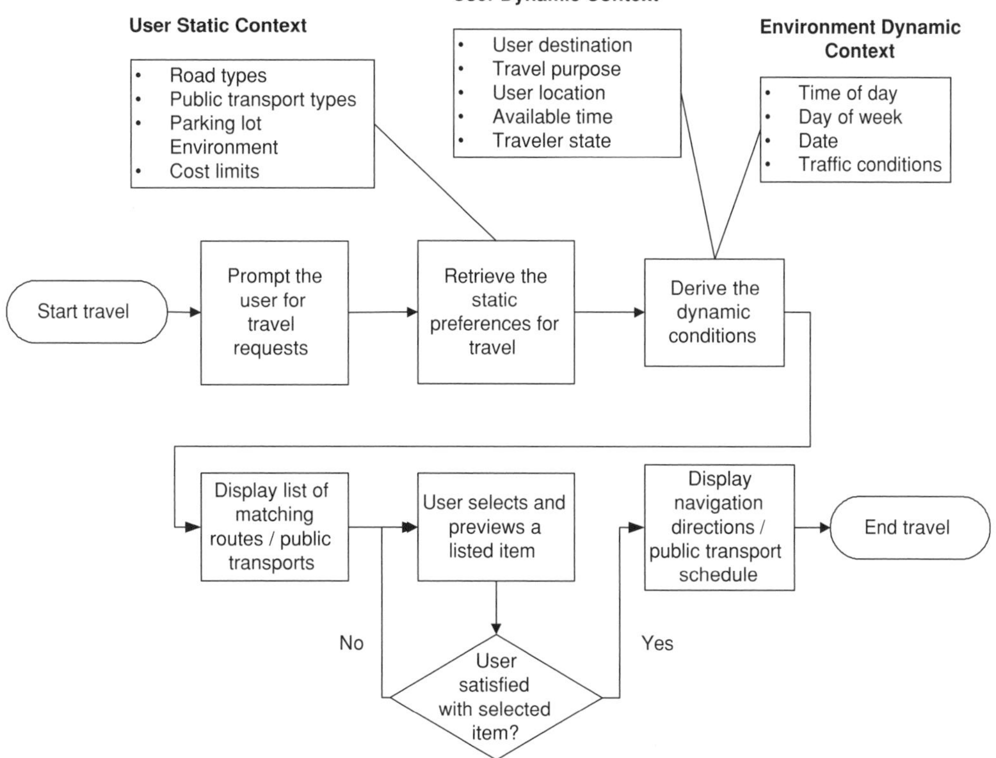

Figure 8.8 Travel application activity diagram with example context elements.

Listing 8.6 shows the associated RDFS code of the user's dynamic context classes and properties.

```
<?xml version="1.0"?>
<rdf:RDF
  xmlns:rdf="http://www.w3.org/1999/02/22-rdf-syntax-ns#"
  xmlns:rdfs="http://www.w3.org/2000/01/rdf-schema#">

<!-- user context hierarchy -->

<rdf:Description rdf:ID="User-Context">
  <rdf:type rdf:resource="http://www.w3.org/2000/01/
    rdf-schema#Class"/>
</rdf:Description>
<rdf:Description rdf:ID="User-Dynamic-Context">
    <rdf:type rdf:resource="http://www.w3.org/2000/01/
      rdf-schema#Class"/>
```

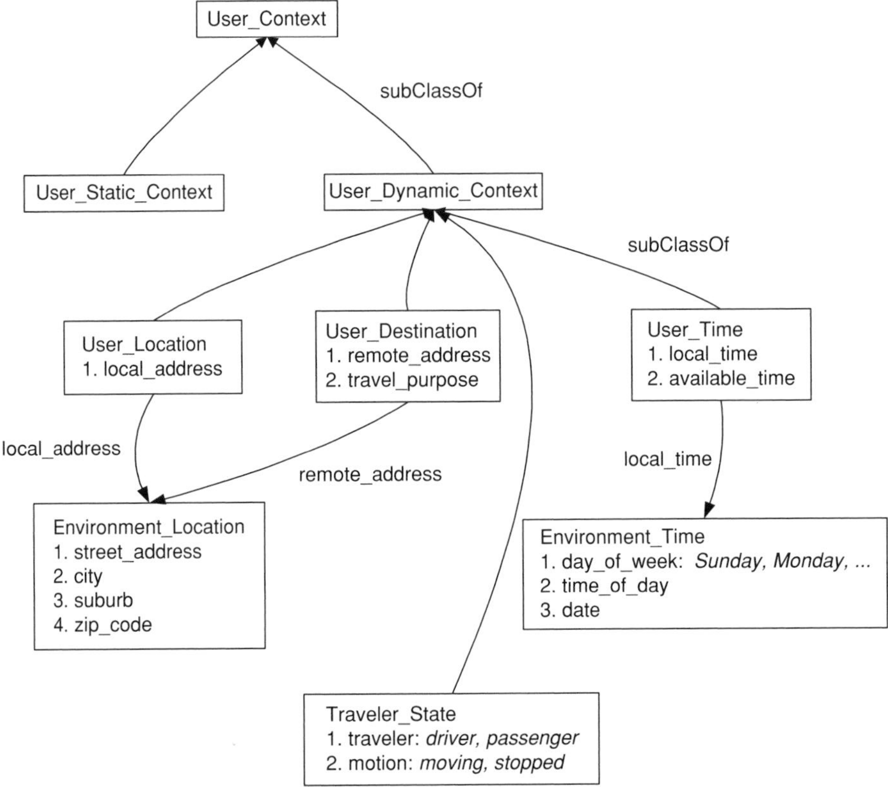

Figure 8.9 RDFS graph of user dynamic context in travel scenarios.

```
   <rdfs:subClassOf rdf:resource="#User-Context"/>
</rdf:Description>
<rdf:Description rdf:ID="User-Static-Context">
   <rdf:type rdf:resource="http://www.w3.org/2000/01/
     rdf-schema#Class"/>
   <rdfs:subClassOf rdf:resource="#User-Context"/>
</rdf:Description>

<!-- user dynamic context classes -->

<rdf:Description rdf:ID="User-Location">
   <rdf:type rdf:resource="http://www.w3.org/2000/01/
     rdf-schema#Class"/>
   <rdfs:subClassOf rdf:resource="#User-Dynamic-Context"/>
</rdf:Description>
<rdf:Description rdf:ID="User-Destination">
```

```
    <rdf:type rdf:resource=“http://www.w3.org/2000/01/
     rdf-schema#Class”/>
    <rdfs:subClassOf rdf:resource=“#User-Dynamic-Context”/>
</rdf:Description>
<rdf:Description rdf:ID=“User-Time”>
    <rdf:type rdf:resource=“http://www.w3.org/2000/01/
     rdf-schema#Class”/>
    <rdfs:subClassOf rdf:resource=“#User-Dynamic-Context”/>
</rdf:Description>
<rdf:Description rdf:ID=“Traveler-State”>
    <rdf:type rdf:resource=“http://www.w3.org/2000/01/
     rdf-schema#Class”/>
    <rdfs:subClassOf rdf:resource=“#User-Dynamic-Context”/>
</rdf:Description>
<!-- properties of user dynamic context classes -->
<rdf:Description rdf:ID=“local-address”>
    <rdf:type rdf:resource=“http://www.w3.org/1999/02/
     22-rdf-syntax-ns#Property”/>
    <rdfs:domain rdf:resource=“#User-Location”/>
    <rdfs:range rdf:resource=“#Environment-Location”/>
</rdf:Description>
<rdf:Description rdf:ID=“remote-address”>
    <rdf:type rdf:resource=“http://www.w3.org/1999/02/
     22-rdf-syntax-ns#Property”/>
    <rdfs:domain rdf:resource=“#User-Destination”/>
    <rdfs:range rdf:resource=“#Environment-Location”/>
</rdf:Description>
<rdf:Description rdf:ID=“travel-purpose”>
    <rdf:type rdf:resource=“http://www.w3.org/1999/02/
     22-rdf-syntax-ns#Property”/>
    <rdfs:domain rdf:resource=“#User-Destination”/>
    <rdfs:range rdf:resource=“http://www.w3.org/2001/
     XMLSchema#string”/>
</rdf:Description>
<rdf:Description rdf:ID=“local-time”>
    <rdf:type rdf:resource=“http://www.w3.org/1999/02/
     22-rdf-syntax-ns#Property”/>
    <rdfs:domain rdf:resource=“#User-Time”/>
    <rdfs:range rdf:resource=“#Environment-Time”/>
```

```
</rdf:Description>
<rdf:Description rdf:ID=“available-time”>
  <rdf:type rdf:resource=“http://www.w3.org/1999/02/
   22-rdf-syntax-ns#Property”/>
  <rdfs:domain rdf:resource=“#User-Time”/>
  <rdfs:range rdf:resource=“http://www.w3.org/2001/
   XMLSchema#duration”/>
</rdf:Description>
<rdf:Description rdf:ID=“traveler”>
  <rdf:type rdf:resource=“http://www.w3.org/1999/02/
   22-rdf-syntax-ns#Property”/>
  <rdfs:domain rdf:resource=“#Traveler-State”/>
  <rdfs:range rdf:resource=“#Traveler-Value”/>
</rdf:Description>
<rdf:Description rdf:ID=“motion”>
  <rdf:type rdf:resource=“http://www.w3.org/1999/02/
   22-rdf-syntax-ns#Property”/>
  <rdfs:domain rdf:resource=“#Traveler-State”/>
  <rdfs:range rdf:resource=“#Motion-Value”/>
</rdf:Description>

</rdf:RDF>
```

Listing 8.6 RDFS code of user dynamic context in travel scenarios

The RDFS graph in Figure 8.10 depicts RDFS resources of user static context in the described travel scenarios.

The associated RDFS code in Listing 8.7 shows the user's static context classes and properties.

```
<?xml version=“1.0”?>
<rdf:RDF
 xmlns:rdf=“http://www.w3.org/1999/02/22-rdf-syntax-ns#”
 xmlns:rdfs=“http://www.w3.org/2000/01/rdf-schema#”>

<!-- user context hierarchy -->

<rdf:Description rdf:ID=“User-Context”>
  <rdf:type rdf:resource=“http://www.w3.org/2000/01/
   rdf-schema#Class”/>
</rdf:Description>
<rdf:Description rdf:ID=“User-Dynamic-Context”>
```

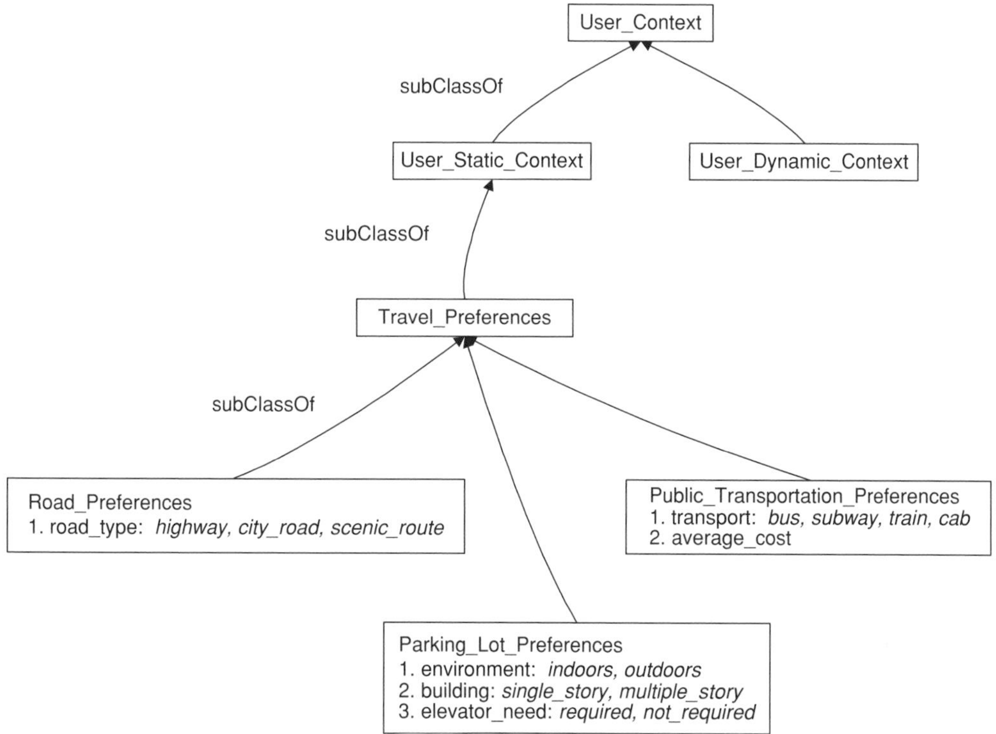

Figure 8.10 RDFS graph of user static context in travel scenarios.

```
    <rdf:type rdf:resource="http://www.w3.org/2000/01/
     rdf-schema#Class"/>
    <rdfs:subClassOf rdf:resource="#User-Context"/>
</rdf:Description>
<rdf:Description rdf:ID="User-Static-Context">
    <rdf:type rdf:resource="http://www.w3.org/2000/01/
     rdf-schema#Class"/>
    <rdfs:subClassOf rdf:resource="#User-Context"/>
</rdf:Description>

<!-- user static context classes -->

<rdf:Description rdf:ID="Travel-Preferences">
    <rdf:type rdf:resource="http://www.w3.org/2000/01/
     rdf-schema#Class"/>
    <rdfs:subClassOf rdf:resource="#User-Static-Context"/>
</rdf:Description>
<rdf:Description rdf:ID="Road-Preferences">
```

```
    <rdf:type rdf:resource="http://www.w3.org/2000/01/
     rdf-schema#Class"/>
    <rdfs:subClassOf rdf:resource="#Travel-Preferences"/>
</rdf:Description>
<rdf:Description rdf:ID=
  "Public-Transportation-Preferences">
    <rdf:type rdf:resource="http://www.w3.org/2000/01/
     rdf-schema#Class"/>
    <rdfs:subClassOf rdf:resource="#Travel-Preferences"/>
</rdf:Description>
<rdf:Description rdf:ID="Parking-Lot-Preferences">
    <rdf:type rdf:resource="http://www.w3.org/2000/01/
     rdf-schema#Class"/>
    <rdfs:subClassOf rdf:resource="#Travel-Preferences"/>
</rdf:Description>

<!-- properties of user static context classes -->

<rdf:Description rdf:ID="road-type">
    <rdf:type rdf:resource="http://www.w3.org/1999/02/
     22-rdf-syntax-ns#Property"/>
    <rdf:type rdf:resource="http://www.w3.org/2000/01/
     rdf-schema#Seq"/>
    <rdfs:domain rdf:resource="#Road-Preferences"/>
    <rdfs:range rdf:resource="#Road-Type-Value"/>
</rdf:Description>
<rdf:Description rdf:ID="transport">
    <rdf:type rdf:resource="http://www.w3.org/1999/02/
     22-rdf-syntax-ns#Property"/>
    <rdf:type rdf:resource="http://www.w3.org/2000/01/
     rdf-schema#Seq"/>
    <rdfs:domain rdf:resource=
     "#Public-Transportation-Preferences"/>
    <rdfs:range rdf:resource="#Transport-Value"/>
</rdf:Description>
<rdf:Description rdf:ID="average-cost">
    <rdf:type rdf:resource="http://www.w3.org/1999/02/
     22-rdf-syntax-ns#Property"/>
    <rdfs:range rdf:resource="http://www.w3.org/2001/
     XMLSchema#decimal"/>
</rdf:Description>
```

```
<rdf:Description rdf:ID="environment">
  <rdf:type rdf:resource="http://www.w3.org/1999/02/
   22-rdf-syntax-ns#Property"/>
  <rdf:type rdf:resource="http://www.w3.org/2000/01/
   rdf-schema#Alt"/>
  <rdfs:range rdf:resource="#Environment_Value"/>
</rdf:Description>
<rdf:Description rdf:ID="building">
  <rdf:type rdf:resource="http://www.w3.org/1999/02/
   22-rdf-syntax-ns#Property"/>
  <rdf:type rdf:resource="http://www.w3.org/2000/01/
   rdf-schema#Alt"/>
  <rdfs:range rdf:resource="#Building_Value"/>
</rdf:Description>
<rdf:Description rdf:ID="elevator_need">
  <rdf:type rdf:resource="http://www.w3.org/1999/02/
   22-rdf-syntax-ns#Property"/>
  <rdf:type rdf:resource="http://www.w3.org/2000/01/
   rdf-schema#Alt"/>
  <rdfs:range rdf:resource="#Elevator_Need_Value"/>
</rdf:Description>

</rdf:RDF>
```

Listing 8.7 RDFS code of user static context in travel scenarios

Possible values for those properties with specified ranges are listed in the RDFS class hierarchy graph in Figure 8.10.

8.5 User tracking and communication scenarios and context

A mobile user can request information about the location of other users with whom he or she is affiliated through membership in a common group. For example, emergency staff personnel may wish to know the location of other staff in their vicinity if they require special support. Similarly, if a family, or for that matter a party of tourists, is visiting a large theme park such as Orlando's Disney World, and different group members are in different sections of the park, they may wish to coordinate their activities for the day. In this case, the family, or party of tourists, is a group whose members wish to track the location of other members. In this scenario, dynamic context consists of the user's location whereas static context consists of the user's association with a group,

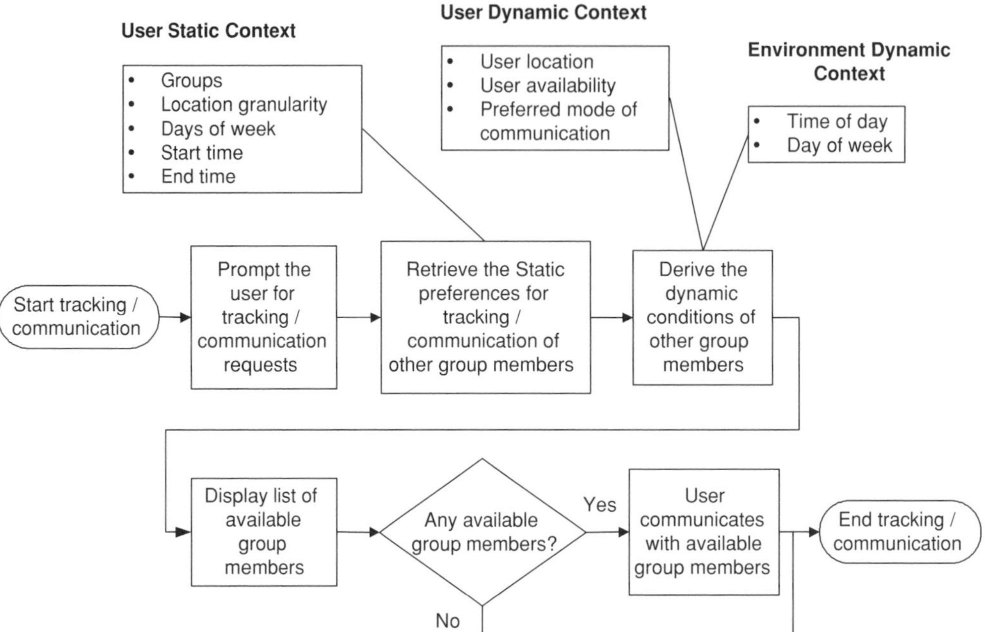

Figure 8.11 Tracking / communication application activity diagram with example context elements.

and any privacy constraints that affect the disclosure of the user's location to other group members.

Mobile users may want to control their disclosed location granularity. For example, users may be willing to inform other members in which town they are but may not want to divulge their precise geographic coordinates. Location disclosure could be time-related since at certain times of the day, or certain days of the week, mobile users may not want other group members to know their whereabouts, and therefore may want to activate corresponding privacy controls. On the other hand, for a family visiting a theme park, parents with children equipped with mobile terminals may want to know the precise location of their children at all times.

Once group members have been located, the mobile user can determine with whom to establish communication. If the group is a work group, communication could be directed to specific group members based on their particular skills and their availability. If the group is a party of tourists, then they may want to communicate with each other to get together for lunch. In both cases, dynamic context elements that indicate availability and preferred mode of communication could be used to determine what communication method would be best. For example, in a theater, considerate mobile users would rather receive text messages, while they could receive voice calls when outdoors. Figure 8.11 shows a travel application's activity diagram for the above scenarios and the associated static and dynamic elements of context.

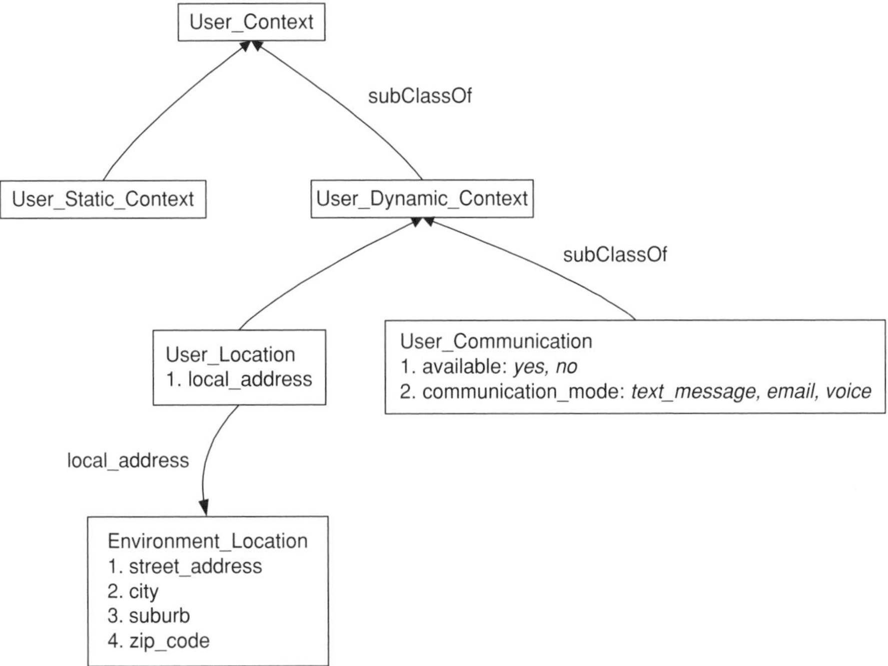

Figure 8.12 RDFS graph of user dynamic context in user tracking and communication scenarios.

The RDFS graph in Figure 8.12 depicts RDFS resources of user dynamic context in the described user tracking and communication scenarios.

Listing 8.8 shows the associated RDFS code of the user's dynamic context classes and properties.

```
<?xml version="1.0"?>
<rdf:RDF
 xmlns:rdf="http://www.w3.org/1999/02/22-rdf-syntax-ns#"
 xmlns:rdfs="http://www.w3.org/2000/01/rdf-schema#">

<!-- user context hierarchy -->

<rdf:Description rdf:ID="User-Context">
  <rdf:type rdf:resource="http://www.w3.org/2000/01/
   rdf-schema#Class"/>
</rdf:Description>
<rdf:Descripotion rdf:ID="User-Dynamic-Context">
   <rdf:type rdf:resource="http://www.w3.org/2000/01/
    rdf-schema#Class"/>
   <rdfs:subClassOf rdf:resource="#User-Context"/>
```

```
</rdf:Description>
<rdf:Description rdf:ID=“User-Static-Context”>
   <rdf:type rdf:resource=“http://www.w3.org/2000/01/
     rdf-schema#Class”/>
   <rdfs:subClassOf rdf:resource=“#User-Context”/>
</rdf:Description>
<!-- user dynamic context classes -->
<rdf:Description rdf:ID=“User-Location”>
   <rdf:type rdf:resource=“http://www.w3.org/2000/01/
     rdf-schema#Class”/>
   <rdfs:subClassOf rdf:resource=“#User-Dynamic-Context”/>
</rdf:Description>
<rdf:Description rdf:ID=“User-Communication”>
   <rdf:type rdf:resource=“http://www.w3.org/2000/01/
     rdf-schema#Class”/>
   <rdfs:subClassOf rdf:resource=“#User-Dynamic-Context”/>
</rdf:Description>
<!-- properties of user dynamic context classes -->
<rdf:Description rdf:ID=“local-address”>
   <rdf:type rdf:resource=“http://www.w3.org/1999/02/
     22-rdf-syntax-ns#Property”/>
   <rdfs:domain rdf:resource=“#User-Location”/>
   <rdfs:range rdf:resource=“#Environment-Location”/>
</rdf:Description>
<rdf:Description rdf:ID=“available”>
   <rdf:type rdf:resource=“http://www.w3.org/1999/02/
     22-rdf-syntax-ns#Property”/>
<rdf:type rdf:resource=“http://www.w3.org/2000/01/
  rdf-schema#Alt”/>
   <rdfs:range rdf:resource=“http://www.w3.org/2001/
     XMLSchema#string”/>
</rdf:Description>
<rdf:Description rdf:ID=“communication-mode”>
   <rdf:type rdf:resource=“http://www.w3.org/1999/02/
     22-rdf-syntax-ns#Property”/>
   <rdf:type rdf:resource=“http://www.w3.org/2000/01/
     rdf-schema#Seq”/>
```

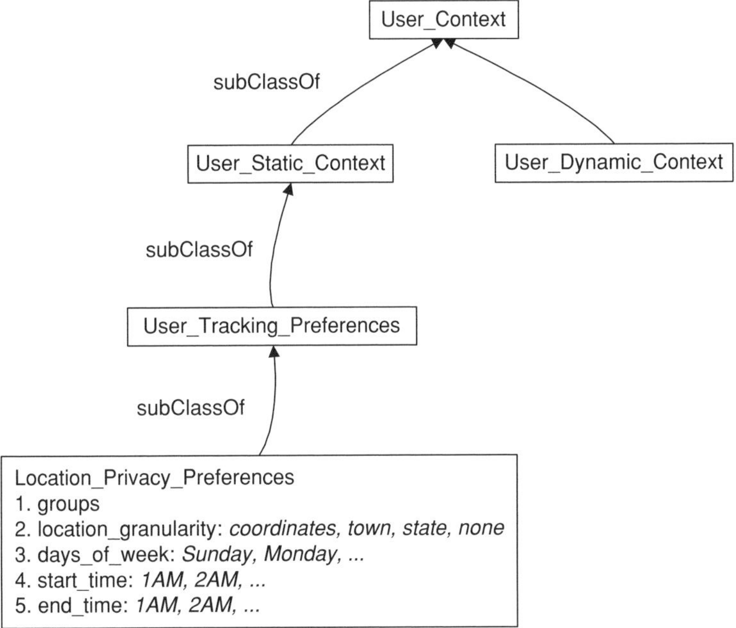

Figure 8.13 RDFS graph of user static context in user tracking and communication scenarios.

```
        <rdfs:domain rdf:resource="#User-Communication"/>
        <rdfs:range rdf:resource="#Communication-Mode-Value"/>
    </rdf:Description>
    </rdf:RDF>
```

Listing 8.8 RDFS code of user dynamic context in user tracking and communication scenarios

The RDFS graph in Figure 8.13 depicts RDFS resources of user static context in the described travel scenarios.

Listing 8.9 shows the associated RDFS code of the user's static context classes and properties.

```
<?xml version="1.0"?>
<rdf:RDF
  xmlns:rdf="http://www.w3.org/1999/02/22-rdf-syntax-ns#"
  xmlns:rdfs="http://www.w3.org/2000/01/rdf-schema#">
<!-- user context hierarchy -->
<rdf:Description rdf:ID="User-Context">
```

```
  <rdf:type rdf:resource="http://www.w3.org/2000/01/
    rdf-schema#Class"/>
</rdf:Description>
<rdf:Description rdf:ID="User-Dynamic-Context">
  <rdf:type rdf:resource="http://www.w3.org/2000/01/
    rdf-schema#Class"/>
  <rdfs:subClassOf rdf:resource="#User-Context"/>
</rdf:Description>
<rdf:Description rdf:ID="User-Static-Context">
  <rdf:type rdf:resource="http://www.w3.org/2000/01/
    rdf-schema#Class"/>
  <rdfs:subClassOf rdf:resource="#User-Context"/>
</rdf:Description>

<!-- user static context classes -->

<rdf:Description rdf:ID="Location-Privacy-Preferences">
  <rdf:type rdf:resource="http://www.w3.org/2000/01/
    rdf-schema#Class"/>
  <rdfs:subClassOf rdf:resource="#User-Static-Context"/>
</rdf:Description>

<!-- properties of user static context classes -->

<rdf:Description rdf:ID="group">
  <rdf:type rdf:resource="http://www.w3.org/1999/02/
    22-rdf-syntax-ns#Property"/>
  <rdfs:domain rdf:resource="#Location-Privacy-Preferences"/>
  <rdfs:range rdf:resource="http://www.w3.org/2001/
    XMLSchema#string"/>
</rdf:Description>
<rdf:Description rdf:ID="location-granularity">
  <rdf:type rdf:resource="http://www.w3.org/1999/02/
    22-rdf-syntax-ns#Property"/>
  <rdf:type rdf:resource="http://www.w3.org/2000/01/
    rdf-schema#Alt"/>
  <rdfs:domain rdf:resource="#Location-Privacy-Preferences"/>
  <rdfs:range rdf:resource="#Location-Granularity-Value"/>
</rdf:Description>
<rdf:Description rdf:ID="days-of-week">
  <rdf:type rdf:resource="http://www.w3.org/1999/02/
```

```
      22-rdf-syntax-ns#Property”/>
    <rdf:type rdf:resource=“http://www.w3.org/2000/01/
      rdf-schema#Bag”/>
    <rdfs:domain rdf:resource=
      “#Location-Privacy-Preferences”/>
    <rdfs:range rdf:resource=“#Days-Of-Week-Value”/>
</rdf:Description>
<rdf:Description rdf:ID=“start-time”>
    <rdf:type rdf:resource=“http://www.w3.org/1999/02/
      22-rdf-syntax-ns#Property”/>
    <rdfs:domain rdf:resource=“#Location-Privacy-Preferences”/>
    <rdfs:range rdf:resource=“http://www.w3.org/2001/
      XMLSchema#time”/>
</rdf:Description>
<rdf:Description rdf:ID=“end-time”>
    <rdf:type rdf:resource=“http://www.w3.org/1999/02/
      22-rdf-syntax-ns#Property”/>
    <rdfs:domain rdf:resource=“#Location-Privacy-Preferences”/>
    <rdfs:range rdf:resource=“http://www.w3.org/2001/
      XMLSchema#time”/>
</rdf:Description>

</rdf:RDF>
```

Listing 8.9 RDFS code of user static context in user tracking and communication scenarios

Possible values for those properties with specified ranges are listed in the RDFS class hierarchy graph in Figure 8.13.

9 XSLT for Web content presentation

With the proliferation of mobile terminal types, their different screen features, and the different supported browsers, new approaches were warranted to support flexible development of screen interfaces. XML-based presentation approaches were developed to cater for this need. In this chapter we describe the different Web content sources and how to convert them to displayable markup. Database content can be represented in XML documents and an XSLT processor can apply XSLT style sheets to these representations to generate markup that can be then rendered by mobile terminal browsers. We describe core capabilities of XSLT style sheet programming and how to leverage XSLT for generating displayable content that reflects context-awareness of a mobile user's situation.

9.1 Mobile Web information content

Web content for mobile services can be generated either from the HTML pages accessed on the wired Internet, or directly from the content databases that store the source data. These latter content databases form what is referred to as the "deep Web" [1], since their content is not directly accessible on the static Web pages that form the "surface Web".

9.1.1 HTML source content

An existing wired Internet site can be converted from HTML to a site that has pages in WML, XHTML, or even HTML, suitable for display on a mobile terminal. When converting an existing HTML site, the developer must first decide what elements are suitable for display on a smaller screen mobile terminal, and then design the smaller page layouts, links, and input forms. A site that has small amounts of text, without much reliance on graphics and image formats, would be suitable for conversion.

Conversion tools are either automated, meaning that they convert any HTML page using preset rules, or configurable, meaning that the user can affect the conversion by

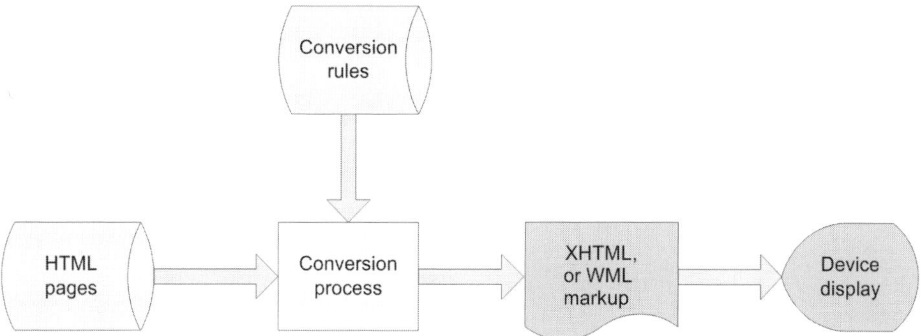

Figure 9.1 HTML content conversion. Source: IEEE © 2003. All rights reserved.

choosing which elements get converted (see Figure 9.1). For example, IBM's WebSphere Transcoding Publisher [2] includes an automated converter tool that will generate one or more WML decks from a given set of HTML pages. This tool can be used either to statically prepare WML content ahead of any demands, or it can be activated dynamically on a per user request. However, since the same conversion rules apply to any HTML page, the generated output might not display the data effectively. Automated conversion is therefore rarely valuable. An improvement to this approach is provided by annotating HTML pages with new tags or comments that can be used to select and tailor content for display on different types of mobile terminals.

The drawback of generating mobile content from HTML pages is that the resultant collection of pages cannot simultaneously satisfy multiple tasks that mobile users may want to accomplish. Since the HTML pages represent static data with no associated semantics, the conversion rules cannot adapt the generated markup (for example, WML) to a user's specific task and associated context.

9.1.2 Database source content

When the source data for generating mobile Web pages is stored in files or in a database, Web programmers usually have more flexibility in choosing what data could be shown on a mobile terminal's screen. This flexibility is preconditioned on the availability of data attributes that profile the source data so that a content selection program can choose content that meets context constraints such as user preferences, location relevance, user situation parameters (the user is driving, the user is in a meeting, etc), and any environment aspects such as time of day, and noise level. The content selection program will generate an intermediary format from the selected content. This format is then converted to displayable markup such as WML, taking into account the mobile terminal's screen characteristics.

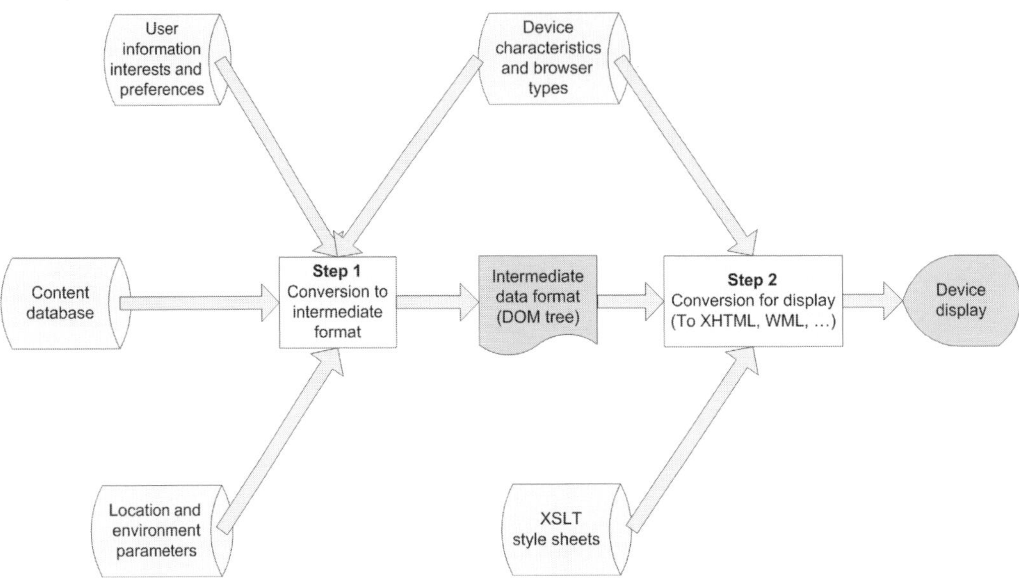

Figure 9.2 Dynamic content generation. Source: IEEE © 2003. All rights reserved.

9.2 Dynamic content generation

In the following sections, we expand on the case when the Web content used to generate displayable content is stored in a database as shown in Figure 9.2.

9.2.1 DOM tree generation

The first step, as Figure 9.2 shows, is to retrieve the requested information from a content database taking into account the context of the user and environment as well as the mobile terminal characteristics. The user's mobile terminal may, for example, support only text and gray scale image display, and a color video should not be retrieved in this case. A database API such as Java Database Connectivity (JDBC) is used to extract the relevant content data, and this data is then converted to a tree representation, referred to as the Document Object Model (DOM) [3].

The W3C DOM is a platform and language-neutral interface that allows programs to represent and dynamically access and update the content, structure and style of valid HTML and XML documents. The DOM API allows a well-formed DOM tree to be built from previously collected data, for example, from the result set retrieved from a database. Listing 9.1 shows an example of an XML representation of the generated DOM for a visitor standing in front of an exhibit of the experimental Web-enabled Motorola Museum [4]. The museum visitor has specified information interests to include history and business items, and has requested that exhibit information media be in audio format.

How does the Web-enabled museum take context data into account when it generates DOM trees? First, the user selects information interests (for example, history) and preferences (for example, audio only) from a list of options displayed on the museum-supplied wireless terminal. The user selections are sent to the museum server and recorded in a user context database. Based on the user's requests for information, exhibit location notifications received by the mobile terminal, the mobile terminal capabilities (for example, screen size), and the prerecorded user interests and preferences, the museum's content generation program dynamically generates JDBC statements that query the museum content database. These queries return result sets that are acted on by the DOM API to create a DOM tree structure whose XML representation is shown in Listing 9.1.

```
<Exhibit id="4">
  <Exhibit_Title>The Company Begins </Exhibit_Title>

  <Audio_Stop>
    <Source href="http://source/museum/audio/Exhibit4.asx">
      <Title>Audio – The Company Begins</Title>
    </Source>
  </Audio_Stop>

  <History>
    <Item Item_id="55415">
      <Audio src="http://source/images/av/winmedia/269_56.asf"/>
      <Item_Title>Paul V. Galvin reflecting on his
                  childhood snack food sales business, 1959.
      </Item_Title>
    </Item>
  </History>

  <Business>
    <Item Item_id="24564">
      <Audio src="http://source/images/av/winmedia/452_56.asf"/>
      <Item_Title>Paul V. Galvin starting his business
      </Item_Title>
    </Item>
  </Business>

</Exhibit>
```

Listing 9.1 XML of the Motorola Museum exhibit DOM tree

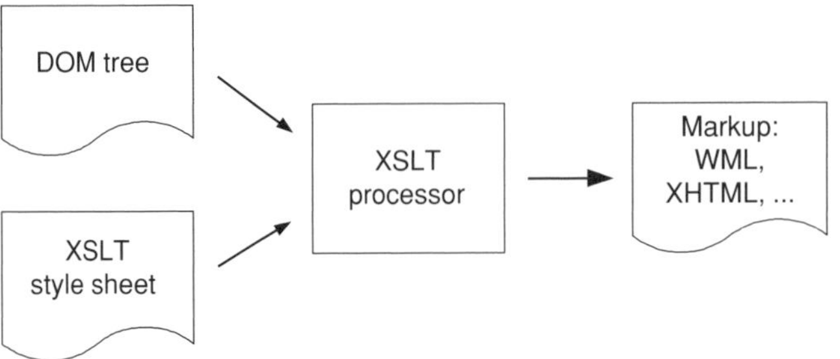

Figure 9.3 Separate DOM content and XSLT presentation generation rules.

9.2.2 Style sheets for output formatting

XML tags allow us to attach semantics to each section of a document, and these sections can be located, extracted, analyzed, and reused. However, DOM content that is structurally tagged with XML is not presentable since the tags do not specify the format to be applied to the text they contain. Formatting information therefore needs to be obtained from another source. Depending on the content types supported by the client browser, an appropriate formatting sheet in Extensible Style Sheet Language Transformations (XSLT) [5] is used to transform the DOM content representation into appropriate markup for display on the user's mobile terminal. Such a formatting sheet is called a "style sheet". For example, if the mobile terminal's browser supports only WML, an XSLT style sheet that generates WML markup will be chosen. In addition, depending on the mobile terminal screen size, the XSLT code decides what information to display on the screen; for example, a large header banner may be omitted from a small screen. These content conversions are represented in the second step in Figure 9.2.

An important benefit of using style sheets is that one style sheet can be substituted for another in order to present the material in a different way. This may be done, for example, to present the same information to a different type of mobile terminal, or to satisfy the needs of a new audience. The program that performs style sheet-based content transformation is referred to as an XSLT processor. Style sheets can be embedded within the XML documents, or can be referenced by including a reference in the XML document, or can be passed to the processor through programmatic selection. Figure 9.3 shows the case when the style sheet is external to the XML document and both files are submitted for processing to the XSLT processor.

Cascading style sheets (CSS), also referred to as wireless CSS (WCSS), are another type of style sheet used for display purposes only. The WAP Forum defined a CSS specification [6], and many latest model phones support CSS. A CSS contains a set of rules, each one specifying how a specific element in the document is to be presented. For example, Listing 9.2 shows how different display attributes of text positioning and font are assigned to XML elements of Listing 9.1.

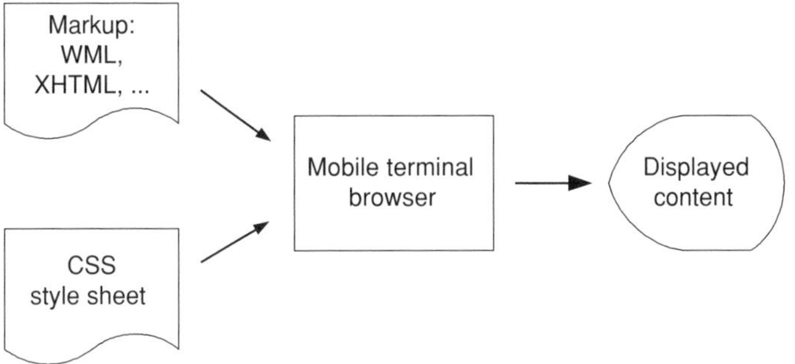

Figure 9.4 Separate markup and CSS presentation directives.

```
[1]   Item_Title {display: inline}
[2]   Exhibit_Title, Title {display: block}
[3]   Exhibit_Title {font-size: 1.3em}
[4]   Title {font-style: italic}
[5]   Item_Title {margin: 0.5em}
```

Listing 9.2 CSS formatting of XML elements

References to WCSS documents are included with *<link>* elements in the markup sent to a mobile terminal, and these external WCSS documents are processed by the mobile terminal's browser in a cascade fashion, that is in their order of appearance. The benefit of WCSS is that content presentation decisions are not left to a browser's implementation, but are rather clearly defined in the WCSS. This enables presentation consistency across terminal types (Figure 9.4).

In the following section we expand on XSLT, a key technology used to generate markup for displaying Web-based content, and we show how context information can be stated in XSLT to affect markup generation.

9.3 XSL and XSLT display transformations

9.3.1 Formatting and tree transformations

The W3C has specified two related standards, XSL and XSLT, for generating displayable content. Extensible Style Sheet Language (XSL) is a specification for describing formatting information [7]. An XSL document contains styling information such as font characteristics, background color, paragraph formats, and page layouts, as well as embedded XML source content. XSL is an XML language, and therefore relatively easy to learn and apply. Listing 9.3 shows an example code fragment in XSL that specifies character formatting instructions of font style, color, and weight, for an embedded sentence.

```
<fo:block font-family="Times" font-size="14pt" font-style="italic">
  <fo:inline color="red">Hello</fo:inline>,
  <fo:inline font-weight="bold">mobile user!</fo:inline>
</fo:block>
```

Listing 9.3 XSL character formatting

The predecessor of the XSL language is the Document Style and Semantics Specification Language (DSSSL) for use with the Standard Generalized Markup Language (SGML), the precursor to XML. In addition to DSSSL, XSL also builds upon the W3C CSS language, usually used in conjunction with HTML or XHTML. Like CSS instructions, XSL instructions provide more expressiveness than the available HTML default tags. However, while CSS format instructions can be in a separate style sheet file, or in a separate section that is embedded in the content file, XSL instructions are embedded with the XML source text. Figure 9.5(a) shows how an XSL formatting processor uses the XSL formatting objects in conjunction with the embedded XML source elements for the purpose of displaying the XML content.

XSL Transformations (XSLT) is a related W3C specification [5] that defines a transformation language for converting XML documents into other formats, for example, into an XSL document as described previously, or HTML, XHTML, or WML, for rendering by a browser. XSLT defines mapping rules, also referred to as tree transformation rules, for converting source XML elements into target output elements. These mapping rules are contained in style sheets. By substituting one style sheet for another, the same content can be presented in a different view. For example, one style sheet can be used for generating a page in a laptop's Web browser, while another can enable the transformation of the same content for display in a cell phone's Web browser (Figure 9.5(b)).

In the following sections, we introduce XSLT transformation rules, their syntax and capabilities. The examples show how to generate information for display on a mobile terminal where the user and environment context affects the choice of what information to display.

9.3.2 XSLT mapping rules

XSLT mapping rules are specified in "templates". For example, the following template matches an XML <*weather*> element and generates a corresponding HTML *paragraph* structure (XSLT tag names are defined in the *xsl* namespace):

```
<xsl:template match="weather">
  <P>
    <xsl:apply-templates/>
  </P>
</xsl:template>
```

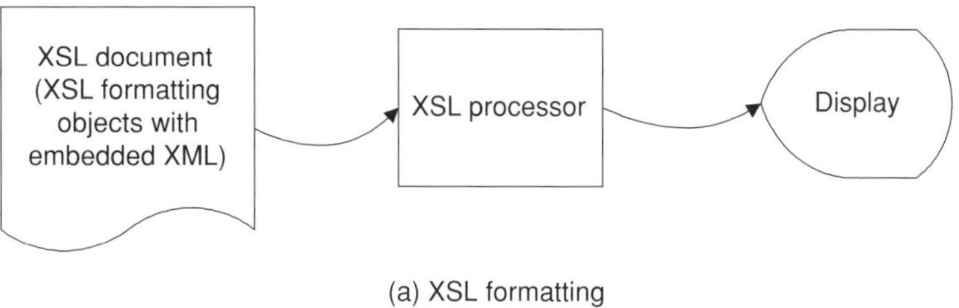

(a) XSL formatting

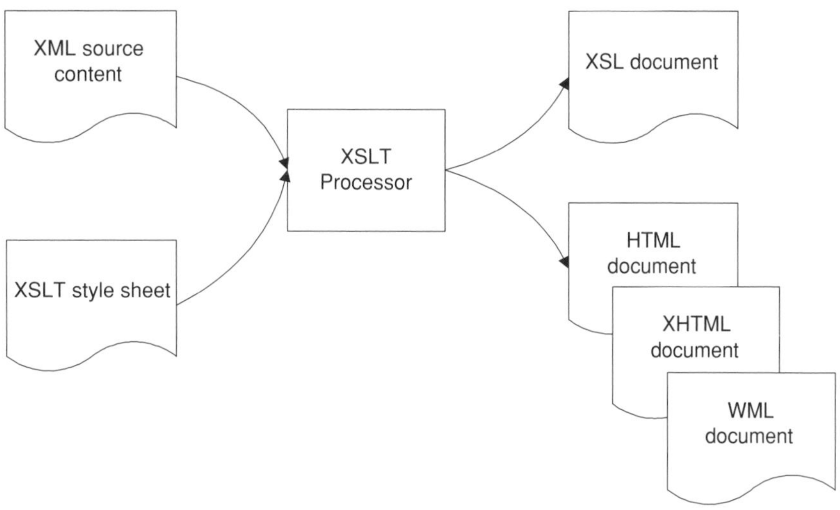

(b) XSLT Tree Transformation

Figure 9.5 XSL processing compared with XSLT processing.

The XML text is

```
<weather> It is windy and rainy </weather>
```

and the generated HTML is

```
<P> It is windy and rainy </P>
```

An XSLT processor performs the processing of a style sheet's templates. The processor will walk through the source XML document, and process templates that match the source elements. The processing starts at the root of the XML tree and progresses through the child elements in what is called "tree-walking". Application of templates is not automatic and needs to be explicitly specified with the *<apply-templates>* element. Each occurrence of an *<apply-templates>* element is an indication to the XSLT processor to process all templates of child elements of the current element. Element text is automatically presented when the element's *<apply-templates>* element is called.

By default, the text of all child elements of the current element will be displayed as well.

How is template processing controlled?

As mentioned previously, the XSLT templates are applied in the order of corresponding XML elements appearance in the source XML tree. The processing order can be altered with the use of a *select* attribute to pick specific elements within the source XML, not necessarily children of the current node. For example, if the the example XSLT template is modified to:

```
<xsl:template match="weather">
  <P>
    <xsl:apply-templates select="tomorrow"/>
  </P>
</xsl:template>
```

and the XML text is

```
<weather>
  <today> It is windy and rainy </today>
  <tomorrow> It will be a sunny day </tomorrow>
</weather>
```

the generated HTML is

```
<P> It will be a sunny day </P>
```

Both texts of the *<today>* and *<tomorrow>* elements would have been presented without the *select* attribute.

To add expressiveness for element selection in XSLT rules, a W3C language specification, XPath [8], comes into play for querying the structure of XML documents. The XPath language is not XML based; however, it is very compact and expressive, so that it is not difficult to understand. XPath expressions can be embedded in XSLT rules for retrieving the source elements to which formatting directives are applied. For example, the previous XSLT rule is expanded in the next example with an XPath expression as the value of the *match* attribute so that it will match only *<city>* elements that have a *weather* attribute value of *today*, and that are contained in *<forecast>* elements:

```
<xsl:template match="forecast//city[@weather='today']">
  <P>
    <xsl:apply-templates/>
  </P>
</xsl:template>
```

While more than one template can refer to the same XML element, only one template can be used for formatting purposes. Typically, the XSLT processor will pick the applicable template closest to the end of the style sheet. To resolve any conflict, a priority attribute can be added to the template, and the XSLT processor selects the template with the highest priority value. Conversely, the same template can apply to multiple XML elements that meet the same match criteria.

9.3.3 Style sheet directed template processing

An alternative approach to determine the order of XML elements processing is to associate identifiers with templates and then use XSLT statements that invoke these templates. Templates can be named and referenced from other templates by including a *name* attribute in the *<template>* element. A named template can then be called using the *<call-template>* element. Furthermore, parameter elements can be defined with the *<param>* element, and an XSLT template can be passed parameter values when called.

This template processing approach can be applied to generate context-aware notifications that can be displayed on a user's mobile terminal. For example, it may be desired to issue a different notification depending on the city where a user is located. The XML source in Listing 9.4 specifies the current town where user Michael is located, and provides notes about points-of-interest in the area's towns.

```
<?xml version="1.0"?>
<pointsOfInterest>

<user name="Michael">
  <currentTown>Highland Park</currentTown>
</user>

<towns>
  <town>
    <name>Highland Park</name>
    <note>Ravinia park is nearby</note>
  </town>
  <town>
    <name>Highland Park</name>
    <note>Botanical Gardens are nearby</note>
  </town>
  <town>
    <name>Evanston</name>
    <note>Century 12 and Cine Arts 6 theaters are nearby</note>
```

```
    </town>
    <town>
      <name>Lincoln Park</name>
      <note>Lincoln Park Zoo is nearby</note>
    </town>
  </towns>

</pointsOfInterest>
```

Listing 9.4 XML source of user location and points of interest

As the user travels between towns, a tourist information service may want to inform the user about nearby points of interest. In Listing 9.4, user Michael is in Highland Park, and the service wishes to issue a corresponding notification by generating the following HTML markup:

```
<P>Ravinia Park is nearby</P>
<P>Botanical Gardens are nearby</P>
```

This can be achieved with the XSL style sheet of Listing 9.5. The *user* template issues a call to the *proximity* template with a parameter whose value is the *<currentTown>* element's value contained within the *<user>* element. Depending on the particular town name, a different notification is output. For user Michael, the XSLT processor finds out that the user is in Highland Park, and it generates the previous notification markup associated with that town.

```
<?xml version="1.0"?>
<xsl:stylesheet xmlns:xsl="http://www.w3.org/1999/XSL/
  Transform" version="1.0">

<xsl:template match="user">
  <xsl:call-template name="proximity">
    <xsl:with-param name="location">
      <xsl:value-of select="currentTown"/>
    </xsl:with-param>
  </xsl:call-template>
</xsl:template>

<xsl:template name="proximity">
  <xsl:param name="location"/>
    <xsl:for-each select="//Towns/Town[name=$location]">
      <P> <xsl:value-of select="note"/> </P>
    </xsl:for-each>
```

```
</xsl:template>

<xsl:template match="//update/towns"></xsl:template>
</xsl:stylesheet>
```

Listing 9.5 XSLT style sheet for generating HTML of points of interest notifications

Looping through multiple elements of the source XML document is enabled with a *<for-each>* element that selects all elements named with the *select* attribute, and applies the enclosed formatting instructions. In the XSL example in Listing 9.5, the list of towns is searched for those that have the same name as the user's current town location. A paragraph structure is generated for each note associated with the current town so that the user can be presented with a list of points of interest.

9.3.4 Determining the output content

The source document element values can be easily transferred to the output document with the help of special purpose elements. As Listing 9.5 shows, the XSLT *<value-of>* element is used to transfer string values of source document elements. The source element is chosen with the *select* attribute. Either an element's string value, or else the string value of an element's attribute can be selected. Attribute names are distinguished from element names by preceding an attribute name with the "@" symbol.

As multiple templates can appear in a style sheet, and a template can be an elaborate construct, there may be situations where it is desired to apply only parts of a template or of the style sheet. The *<if>* and *<choose>* elements enable such selections. The *<if>* element appears jointly with a *test* attribute whose value must be true for processing the template fragment within the element. The *<if>* element could be used, for example, to select points of interest information that is relevant to a particular season. The points of interest source from Listing 9.4 can be augmented with season information as shown in the following XML fragment where each *<note>* element of Highland Park has an added season attribute:

```
<town>
  <name>Highland Park</name>
  <note season="summer">Ravinia park is nearby</note>
</town>
<town>
  <name>Highland Park</name>
  <note season="spring">Botanical Gardens are nearby</note>
</town>
```

The XSL *proximity* template from Listing 9.5 is modified to the following code:

```
<xsl:template name="proximity">
  <xsl:param name="location"/>
    <xsl:for-each select="//towns/town[name=$location]/note">
      <xsl:if test="@season='spring'">
      <P>
        <xsl:value-of select="."/>
      </P>
      </xsl:if>
    </xsl:for-each>
 </xsl:template>
```

The *<if>* element checks whether the *season* attribute of the *<note>* element matches the value *spring*, and only then selects the corresponding notification text. The text is selected from the currenty processed element which is denoted by the character “ . ”. The generated HTML is now:

```
<P> Botanical Gardens are nearby </P>
```

Another way to select options is with the *<choose>* element. This latter element enables the selection of an alternative among a set of elements, where each element has a *<when>* element with an associated *test* attribute similar to the *<if>* element case. For example, the points of interest source from Listing 9.4 can be augmented with entrance cost information as shown in the following XML fragment on Highland Park:

```
<town>
  <name>Highland Park</name>
  <note entranceCost="12.00">Ravinia park is nearby</note>
</town>
<town>
  <name>Highland Park</name>
  <note entranceCost="7.00">Botanical Gardens are nearby</note>
</town>
```

The XSL *proximity* template from Listing 9.5 is modified to the following code:

```
<xsl:template name="proximity">
  <xsl:param name="location"/>
    <xsl:for-each select="//towns/town[name=$location]/note">
      <xsl:choose>
        <xsl:when test="@entranceCost &lt; 10.01">
          <P>Low cost entrance: <xsl:value-of select="."/></P>
        </xsl:when>
        <xsl:when test="@entranceCost &gt; 10.00">
```

```
            <P>High cost entrance: <xsl:value-of select="."/></P>
          </xsl:when>
        </xsl:choose>
      </xsl:for-each>
    </xsl:template>
```

The generated HTML is:

```
<P>High cost entrance: Ravinia park is nearby</P>
<P>Low cost entrance: Botanical Gardens are nearby</P>
```

The previous examples can be extended to handle not just points of interest that are relevant to a user's location context, but also to take account as the user's transportation means. Driving users want to be informed about parking lot locations, and if walking, they may want directions to the train station. Similarly to the case where a user's location was represented with a *<currentTown>* element, the user's transportation context can be expressed in a *<transportationStatus>* element that indicates whether the user is driving or walking. Depending on the user's transportation status, a different indication will be displayed on the user's mobile terminal screen.

Listing 9.6 shows the XML source that contains user Michael's context and towns' context information. Michael's context includes the part of town he is in, as well as his current transportation means. The town context includes parking lot locations for drivers and public transportation station locations for pedestrians.

```
<? xml version="1.0"?>
<pointsOfInterest>

<user name="Michael">
  <currentTown>Evanston</currentTown>
  <currentPart>east</currentPart>
  <currentTransport>pedestrian</currentTransport>
</user>

 <towns>
   <town>
     <name>Evanston</name>
     <area>center</area>
     <driver>Parking is at the corner of Maple and
       Clark</driver>
     <pedestrian>Train station is at the corner of Emerson and
       Ridge</pedestrian>
   </town>
   <town>
```

```
      <name>Evanston</name>
      <area>east</area>
      <driver>Parking is at the corner of Noyes and
       Sheridan</driver>
      <pedestrian>Bus station is at the corner of Central and
       Sheridan</pedestrian>
    </town>
  </towns>

  </pointsOfInterest>
```

Listing 9.6 XML source of user and town information for drivers and pedestrians

The following XSLT template caters to drivers and pedestrian by generating the HTML that provides town information relevant to the user’s transportation means.

```
<?xml version=“1.0”?>
<xsl:stylesheet xmlns:xsl=“http://www.w3.org/1999/XSL/Transform”
 version=“1.0”>

<xsl:template match=“//update/user”>
  <xsl:call-template name=“proximity”>
    <xsl:with-param name=“location”>
      <xsl:value-of select=“currentTown”/>
    </xsl:with-param>
    <xsl:with-param name=“part”>
       <xsl:value-of select=“currentPart”/>
     </xsl:with-param>
     <xsl:with-param name=“transport”>
       <xsl:value-of select=“currentTransport”/>
     </xsl:with-param>
   </xsl:call-template>
 </xsl:template>

 <xsl:template name=“proximity”>
   <xsl:param name=“location”/>
   <xsl:param name=“part”/>
   <xsl:param name=“transport”/>
     <xsl:for-each select=“//towns/town[name=$location]
      [area=$part]”>
       <xsl:choose>
```

```
        <xsl:when test="$transport='driver'">
        <P><xsl:value-of select="driver"/></P>
        </xsl:when>
        <xsl:when test="$transport='pedestrian'">
          <P><xsl:value-of select="pedestrian"/></P>
        </xsl:when>
      </xsl:choose>
    </xsl:for-each>
</xsl:template>

<xsl:template match="//update/towns"></xsl:template>

</xsl:stylesheet>
```

The generated HTML for user Michael is:

```
<P>Bus station is at the corner of Central and Sheridan</P>
```

Listing 9.7 XSLT style sheet for generating HTML of town transportation information

9.3.5 Sorting the output elements

Since an XSLT processor processes the XML source elements in their order of appearance, this is the default order of elements in the output document. XSLT provides a sorting mechanism that allows sorting the elements by the lexicographic order of their element values. This is achieved by adding a *<sort>* empty element in conjunction with the *<apply-templates>* element. In this case, number values appear first, followed by lower-case letter values, and then upper-case letter values.

Rather than using string values to order elements, these can be ordered by their numerical value if so desired. Sort criteria can include specific elements, chosen with the *select* attribute, and multiple sort criteria can be specified simultaneously. The sort elements are applied in order of appearance. For example, if the template is:

```
<xsl:template match="costs">
  <xsl:for-each select="item">
    <xsl:sort select="color"/>
    <xsl:sort data-type="number"/>
    <P> <xsl:apply-templates/> </P>
  </xsl:for-each>
</xsl:template>
```

and the XML text is:

```
<costs>
  <item> 23 <color> yellow </color> </item>
  <item> 91 <color> red </color> </item>
  <item> 13 <color> red </color> </item>
  <item> 17 <color> green </color> </item>
</costs>
```

then the generated HTML is:

```
<P> 17 green </P>
<P> 13 red </P>
<P> 91 red </P>
<P> 23 yellow </P>
```

In this case, the list items were sorted first on the string value of the *<color>* element, and then on the number value within each *<item>* element. The assignment of a *number* value to the *data-type* attribute in the second *<sort>* element ensures that the second sorting is done on number values rather than on lexicographic values.

Including the *<number>* element in the sorted element template, with a *position()* value that indicates the position of the sorted element, can add numbering to the sorted elements. For example, in the *costs* template, the *<number>* element is added to the sorted element to produce the new element:

```
<P><xsl:number value="position()"/> <xsl:apply-templates/></P>
```

The generated HTML now enables the display of a numbered list:

```
<P>1 17 green </P>
<P>2 13 red </P>
<P>3 91 red </P>
<P>4 23 yellow </P>
```

Sorting can be applied to attribute values instead of element values. As we showed previously, we can add the cost of entrance to a point of interest with a corresponding attribute. The following XML fragment of the town of Highland Park includes the entrance cost for each point of interest:

```
<town>
  <name>Highland Park</name>
  <note entranceCost="12.00">Ravinia park is nearby</note>
</town>
<town>
  <name>Highland Park</name>
  <note entranceCost="7.00">Botanical Gardens are nearby</note>
</town>
```

To sort the list of points of interest that are displayed to a mobile user in ascending order of entrance cost, we can modify the XSL *proximity* template from Listing 9.5 to the following code:

```
<xsl:template name="proximity">
  <xsl:param name="location"/>
  <P>In ascending order of entrance cost:</P>
  <xsl:for-each select="//towns/town[name=$location]/ note">
    <xsl:sort select="@entranceCost" data-type= "number"/>
    <P><xsl:value-of select="."/> Cost: <xsl:value-of
          select="@entranceCost"/></P>
  </xsl:for-each>
</xsl:template>
```

The *<sort>* element sorts the points of interest on the numerical value of the *entrance-Cost* attribute. The generated HTML reflects the sorted list, and includes the actual entrance cost as well:

```
<P>In ascending order of entrance cost:</P>
<P>Botanical Gardens are nearby Cost:7.00</P>
<P>Ravinia park is nearby Cost: 12.00</P>
```

REFERENCES AND FURTHER READING

[1] M. K. Bergman, The deep web: Surfacing hidden value. *The Journal of Electronic Publishing*, University of Michigan (Jul. 2001), http://www.press.umich.edu/jep/07-01/bergman.html.

[2] IBM, *Websphere Transcoding Publisher*, http://www-306.ibm.com/software/pervasive/transcoding_publisher/.

[3] A. Le Hors *et al., Document Object Model, Level 3 Core Specification v1.0.* W3C Recommendation (Jun. 2003), http://www.w3.org/TR/2003/WD-DOM-Level-3-Core-20030609/.

[4] A. Pashtan *et al.*, Adapting content for wireless Web Services. *IEEE Internet Computing* (Sep-Oct. 2003), 79–85.

[5] J. Clark, *XSL Transformations (XSLT) vl.0.* W3C Recommendation (Nov. 16, 1999), http:// www.w3.org/TR/xslt.

[6] WAP Forum, *WAP CSS Specification.* WAP-239-WCSS-20011026-a. (http://www.openmobile-alliance.org/tech/affiliates/wap/wapindex.html, Oct. 26, 2001).

[7] S. Adler *et al., Extensible Style Sheet Language (XSL) vl.0.* W3C Recommendation (Oct. 15, 2001), http://www.w3.org/TR/xsl/.

[8] J. Clark and S. DeRose, *XML Path Language (XPath) vl.0.* W3C Recommendation (Nov. 16, 1999), http://www.w3.org/TR/xpath.

N. Bradley, *The XSL Companion* (Addison-Wesley, 2000).

10 Mobile Web network

In this chapter we describe the architecture of a network that enables the delivery of mobile Web information. The network elements include the user's mobile terminal, an application server, a context manager, a service directory, Web services, an authorization, authentication, and accounting (AAA) server, and a wireless gateway that links the wired Web with the mobile terminals. We elaborate upon each network element function and corresponding software infrastructure support. The data content flow and the message flow of an information request scenario are outlined.

A description of the network communication protocols for message exchange follows. We first describe the W3C SOAP protocol for requesting services from remote network servers and provide examples of SOAP request and response messages. We then proceed with a description of the W3C WSDL formal XML representation of a Web service interface, and show examples of WSDL interface descriptions. Finally, we describe the communication protocols used between mobile terminals and network servers.

10.1 Network servers

Wireless Internet services employ network servers, also referred to as application servers, to execute the service business logic and host the databases that store Web content. Initially Web service models included just Web servers that mediated between the client terminal and the database and provided static content in the form of HTML pages. Network servers can complement a Web server, and provide security functions, authenticate users, interact with legacy applications and databases, and generate dynamic content tailored to user requests, their preferences, and client terminal limitations. The dynamic content will contain markup language tags such as XHTML or WML tags that can be interpreted by the mobile terminal's Web browser so that the displayed content appears as Web pages.

Typical services provided by a network server's infrastructure include multiprocessing to handle multiple client requests and simultaneous backend database queries, client session management, page caching, and data streaming. Session management in particular is important for mobile users that have to perform multiple related interactions for a specific task. For example, mobile shopping will require an ecommerce server to track the user's selected items in a virtual shopping cart for later purchase. Similarly, displayed travel directions could be related to previously specified source and destination locations.

In addition to providing a platform for application deployment, network servers provide application development frameworks that include application programming interfaces (APIs) and software development kits (SDKs). For example, Sun's iPlanet server provides a comprehensive workspace for developing multi-tier applications that separate presentation, business, and data logic [1]. iPlanet is built on Java technology in compliance with Sun's Java 2 Platform, Enterprise Edition (J2EE). The J2EE platform includes Enterprise JavaBeans for implementing business logic and Java Server Pages (JSP) for dynamic generation of presentation pages.

An application on a J2EE server is typically partitioned between a few components that include servlets, JSP pages, and Enterprise JavaBeans (EJBs). Servlets are Java components that are addressable via URLs and whose function is to control the application's flow following a user invocation. The servlet processes form input, accesses EJBs that contain business logic, and formats page output by calling JSP pages. Servlets, JSP pages, and EJBs execute within a run-time execution environment referred to as a container.

JSP pages are HTML or XHTML Web pages that reside on the server side and can optionally contain Java code. JSP pages contain JSP-specific tags to perform actions that generate dynamic content. For example, a JSP expression tag can contain Java code that calls a Java method whose returned value is embedded into the page. JSP tags are also used to invoke EJBs when the business logic is contained in external EJBs. The JSP-generated dynamic content forms a display page that is returned to the caller. An alternative to using JSP pages for display generation is to draw on an XSLT translation engine and associated XSL style sheets that are applied to XML content for dynamic page generation.

EJBs provide server-side services such as database access, writing to log files, and client session tracking. There are two kinds of EJB components: entity beans and session beans. Entity beans can be shared by multiple clients; they represent persistent data in a database and support corresponding transactions. Session beans support a single client and are relatively short-lived. For example, a session bean can be created to represent an ecommerce shopping cart that holds a user's purchases. Session beans support specific business logic methods that are invoked by the client application.

Figure 10.1 shows example interactions between a network server's J2EE application components. The user invokes the network server's servlet by submitting a form from the terminal's browser. The servlet then invokes session beans to perform the application's business logic. Session beans, in turn, can invoke entity beans to retrieve or store data. Once the EJBs have returned the requested content, the servlet creates a response page

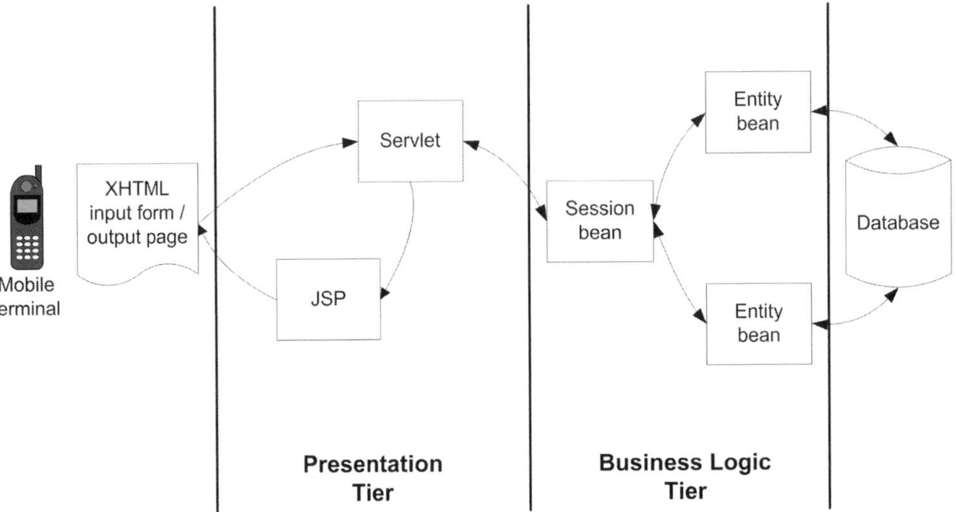

Figure 10.1 Network server J2EE application components.

by forwarding the generated content to a JSP page. The JSP builds the response page that is returned to the user's terminal (Figure 10.1).

10.2 Network architecture

The mobile Web network described here is infrastructure-centric, meaning that most support functions are hosted in network infrastructure servers and the mobile terminals are used mainly for display purposes. The term network infrastructure refers here to both the mobile operator's network of servers and the larger Internet with its associated servers. An infrastructure-centric design approach ensures that the same Web network will be available to serve a broader population of mobile terminals, although it is not precluded that some support functions could be migrated to the terminal.

Whether the mobile user is connected to a wireless wide area network (WWAN) such as a cellular network or to a wireless local area network (WLAN), the network infrastructure requires some key network elements to service mobile user requests for Web-based information. While function allocation to network servers is arbitrary, and in some cases functions can be co-located on the same server, we have chosen an allocation where each server is responsible for a distinct function for a clearer depiction of the main architecture components (see Figure 10.2). Some servers, for example, the AAA server are typically part of the mobile operator's network. Other servers such as a Web service server could be under the administrative domain of external parties. The network element functions are elaborated upon in the following sections.

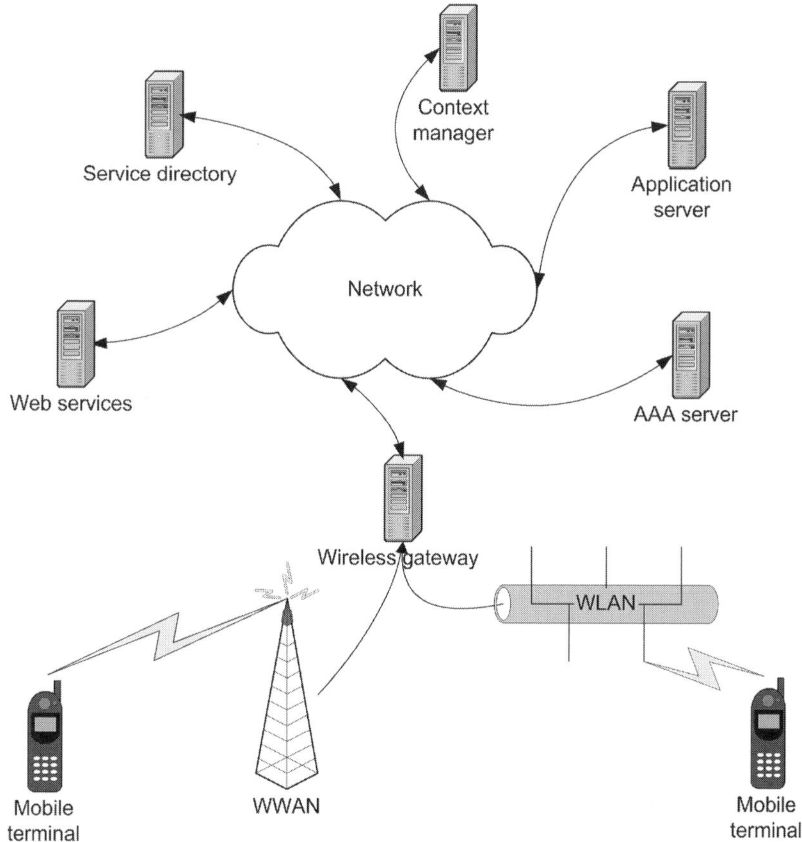

Figure 10.2 Mobile Web services network.

10.2.1 Mobile terminal

The mobile terminals are thin-client browser devices that can display Web-based information when they connect to a WWAN or to a WLAN (Figure 10.2). Mobile terminals that also support application hosting environments, such as Sun's Java 2 Platform, Micro-Edition (J2ME), allow the development of extensive functionality. However, for the purpose of displaying Web content, all content preparation can be performed on the network side. When compared to a mobile terminal, network servers have more powerful processing capabilities, can readily access multiple network databases, and have powerful search and filter capabilities for content generation.

Nevertheless, some support functions, beyond the browser, are required in the mobile terminal. These functions include the ability to control and process data from attached sensors. For example, the mobile terminal's location can be tracked by the terminal itself when equipped with a GPS receiver. Position data is usually collected on a periodic basis, although it could be requested on an as needed basis determined by the serving application. The Context Manager, being the storage server for user context, is the end-recipient of position data.

10.2.2 Application server

The mobile Web network has application logic that is partitioned in this architecture between the Application server front-end and the Context Manager back-end. The Application server hosts the applications that interact with the user's mobile terminal. As the network's front-end, the Application server is responsible for registering users by collecting user preferences in Web forms and forwarding this information to the Context Manager for storage in user profiles.

On receipt of user requests for information, the Application server first authenticates the user by sending appropriate queries to the AAA server. After a user is authenticated, the information request is forwarded to the Context Manager. When the information is returned, the Application server generates appropriate markup, typically by invoking an XSLT processor and submitting to it both the information in XML form and a Web service specific style sheet.

The purpose of the style sheets is to layout the content in a way that is appropriate to the type of requested information. For example, if the content consists of a list of recommended points of interest, such as restaurants or hotels, the style sheet rank-orders the list in a way that is attuned to the user's preferences. In addition, the selected style sheet is specific to a user's mobile terminal browser so that it can generate the appropriate markup.

10.2.3 Context manager

The Context Manager hosts the back-end application logic of the mobile Web network. Mobile user requests for information are forwarded by the Application server to the Context Manager which, in turn, queries Web Services for the requested information. On receipt of the data, the Context Manager filters it according to user preferences expressed in user profiles.

The Context Manager contains a database management system where the user profiles are maintained. Both static user preferences and dynamic context, such as user location, are stored in these user profiles. Context refers to elements such as:

- Tasks that the user wishes to accomplish.
- Absolute location.
- Relative location (to landmarks).
- Physical conditions of the environment (noise, brightness, etc.).
- Movement relative to surroundings (for example, speed, vehicle type).
- Proximity to other users.
- User personal profile and habits.

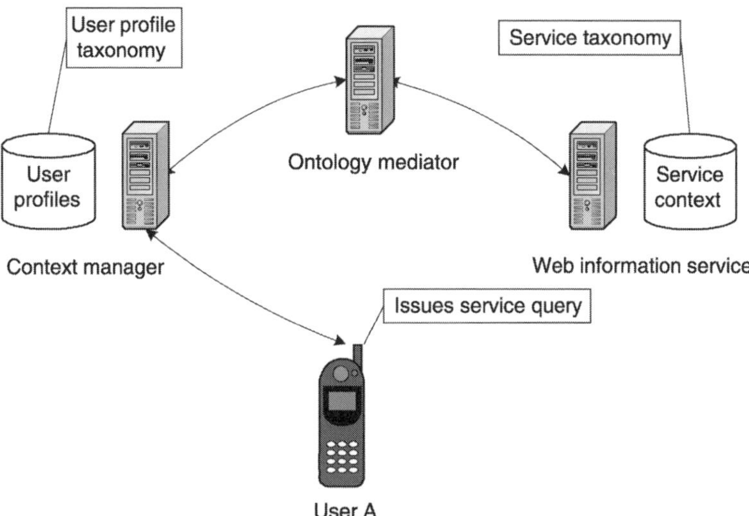

Figure 10.3 Ontology mediator for matching context terms.

- Context history (for conjecturing on the next context).
- Mobile terminal features.

Some of these context elements are specified by the mobile user, for example, the tasks that the user wishes to accomplish. Other context elements require sensors in the mobile terminal or network infrastructure; for example, a GPS receiver in the mobile terminal can provide absolute location, while the infrastructure can measure relative location such as proximity to a park's attraction, or to a store in a shopping mall.

The user and environment context elements determine what specific Web content should be displayed to a mobile user. For example, if a mobile user has issued a restaurant find query, a determination of which restaurants to display can be made based on a comparison of the user's profile-stored preferences with the restaurant's attributes. The format of the delivered information can also be dependent on the user's context. For example, if a user happens to be driving a car he or she may prefer to receive a voice message, however if the user is using public transportation it may be preferable for the information to be sent in the form of a text display.

The Context Manager relies on an Ontology Mediator that can relate terms used in the user profiles with these of the Web services to find the relevant content items that could be displayed on a user's terminal. The Ontology Mediator could be a sub-component of the Context Manager, or it could be a separate server that provides mediation services (see Figure 10.3).

Push capabilities

Not all mobile services are client driven. A mobile user's context may determine that a certain service needs to be activated without explicit user intervention. For example, the

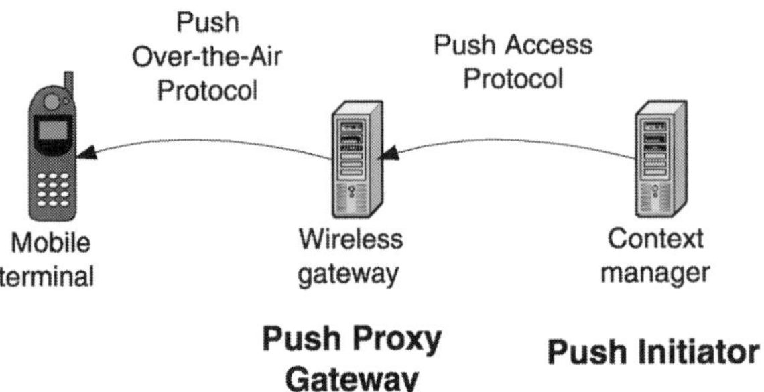

Figure 10.4 Content push services.

fact that a mobile user passes by a movie theater may activate a service that informs the user about a movie he or she previously wanted to see that is playing in the theater. In this example, the combination of a user's location, day of the week, time of day, and the user's movie preferences, causes a Movie service to send a corresponding notification to the user. Similarly, if the Context Manager detects that the mobile user is in the vicinity of a preferred store, and the store happens to carry a special sale event, then a Shopping service could send a corresponding notification to the user.

To support these above services, the Context Manager includes "Push" capabilities that send network originated messages to a mobile client. The WAP Forum has defined the support needed for Push in the form of a Push Initiator (PI) [2]. Since context conditions determine when a message should be pushed to a user, the PI could be hosted in a Context Manager. The pushed message is sent to a Push Proxy Gateway (PPG), typically hosted on a wireless gateway, using the Push Access Protocol (PAP) [3], [4]. If the HTTP protocol is used between the Context Manager and the PPG, and usually this is the case, then the push messages are encapsulated in HTTP POST requests. The PPG then delivers the push content to the receiving mobile terminal with the Push Over-the-Air protocol [5] (Figure 10.4).

The content types delivered by the Context Manager in PAP messages include Service Indication (SI) and Service Loading (SL). An SI usually contains a short text message and a URI providing a service address. The message is presented to the mobile user, who is given the choice to either start the service indicated by the URI immediately, or postpone the SI for later handling. On the other hand, the SL message consists of a URI that is automatically invoked upon receipt in the terminal, that is, the user is not given a choice of activating the service at a later point in time. For most cases, the content pushed to a user's terminal will be of the SI type. SI is the only message type supported by many terminal browsers that provide push support. This is least intrusive, and does not interrupt the user in the midst of other previously engaged wireless transactions.

In the WAP model, the PPG maintains descriptions of the capabilities of the mobile terminals under its control. A requesting Context Manager, can query a PPG with the Client Capabilities Query (CCQ) message for the user's mobile terminal characteristics.

Based on the CCQ response the client can customize the pushed content and adapt it to the terminal's limitations, for example, screen size and graphics support. Since CCQ is an optional feature in the OMA's Push specification, it may not be supported in all deployed PPGs.

10.2.4 Service directory

The Service Directory enables the registration and discovery of Web services. A client application can search the Service Directory, and the returned address of a found service is a URL to which service requests can be submitted. The Context Manager accesses the Service Directory to find appropriate services that are relevant to a mobile user originated request.

Each Service Directory entry can include a list of context attributes that detail service operational characteristics. These service context attributes could provide information about:

- Categories of information provided by the service.
- Geographic vicinity (for example, a store in a shopping mall, or a neighborhood) where the service is applicable.
- Any requirements for co-located users for service interaction (for example, if it's a game played in a park site).
- Any time requirements to listen to or view the information provided by the service (for example, the information could be a short audio, or an expanded text description).
- Effectiveness criteria for different user types: for a user driving in a car, for a user sitting in a train, or for a pedestrian.
- Time-of-day impact on the service. The service provided information may depend on the time of day of the request. For example, different tours may be suggested in a city, depending on the time of day of the request for tour information.
- Service cost (a service could be categorized as expensive, moderate, or free).
- Any mobile terminal requirements (for example, a service may require a color screen of a minimum size).

As can be seen from the above attributes, we extend here the typical notion of service directories to include information that describes under what conditions a service is applicable. When a mobile user attempts to discover a service that he or she can use, additional information of the user's current context needs to be taken into account to find a suitable service candidate.

The OASIS-defined Universal Description, Discovery and Integration (UDDI) directory service [6] can serve in the role of a business directory register in a mobile Web services network. This directory enables the discovery of businesses or organizations, the Web services they make available, and the technical interfaces which may be used to access those services. UDDI can be used for both publicly available services and services only exposed internally within an organization.

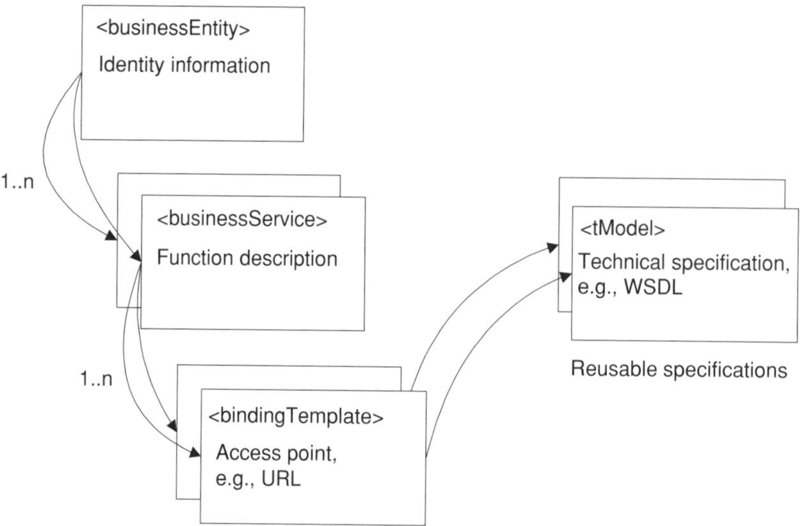

Figure 10.5 The core UDDI information elements.

The UDDI specification [7] presents an information model composed of XML entities. The model's core elements are described in the following and shown in Figure 10.5:

- *<businessEntity>*: describes a business that typically provides Web services. The description includes names, contact information and classification information.
- *<businessService>*: describes a collection of related Web services offered by an organization described by a *<businessEntity>*. Here too, names, descriptions and classification information are included to outline the purpose of the individual Web services.
- *<bindingTemplate>*: describes the technical information necessary to bind and interact with the Web service being described. It must contain either the access point for a given service or an indirection mechanism to reach the access point.
- *<tModel>*: describes a technical model specification representing a concept such as a Web service type, a transport, or a protocol used by Web services. A Web service can point to multiple *<tModel>* elements, and the references to these models are provided by the Web service *<bindingTemplate>*. The *<tModel>* elements themselves are not part of the business information hierarchy, and can be reused between different businesses. For example, a *<tModel>* can contain a URL to specifics about the context of a service that specifies in what user circumstances the service is applicable, and the same *<tModel>* can be used by multiple services that share the same behavior.

In addition to the information model, the UDDI specification defines a programmatic access that includes an inquiry and publisher API, used respectively for searching and registering businesses. Access to these APIs is done with the SOAP protocol over HTTP.

In the next chapter we describe a mobile Web network for tourism services called CATIS. One of the CATIS components is a UDDI service directory that stores Web service references in the form of *<bindingTemplate>* elements that each contain both a service URI address and a reference to a *<tModel>* that identifies the URL of the corresponding WSDL service interface specification.

10.2.5 Web services

A Web Service provides the content that the mobile user is seeking. Typically, each service would be hosted on a separate platform that includes a database server for content storage, and is owned and operated by a corresponding content provider. Example services are a restaurant finder, a hotel finder, and an entertainment park locator.

Database servers provide access to the Web content. When the data is static, for example, in the form of HTML pages, access to the database server is done via a Web server that interprets URLs sent by the mobile terminal to locate the displayable pages. If the data is dynamic, then an additional program is required to issue the SQL queries to the database server, collect the query results, and forward these for further content processing. In this scenario, the URL sent by the mobile terminal specifies the name of the invoked program which could be resident in a separate network server. The following frameworks support the implementation of applications that generate dynamic data.

CGI applications for database access

The Common Gateway Interface (CGI) was the first developed framework that enabled dynamic content generation. CGI provides a portable environment for developing applications since it defines a standard interface between the Web server and the invoked application. The Web server will forward to the CGI application any data that is sent in the client's request and add an indication of both the content's byte length, and the content's type. The CGI application can be written in a variety of programming languages (Perl, Java, etc) and can return either a full document or a reference to one. CGI provides a header in the output returned to the Web server that specifies server directives. These directives indicate the content type, the document's location (if a reference is returned), and a three-digit status code. This header is ASCII text, consisting of lines separated by either linefeeds or carriage returns (or both) followed by a single blank line. The output body then follows in any native format.

When returning a full document, the CGI application needs to specify the MIME type of the document. For example, to send back HTML the header will be: *Content-type: text/html*. When returning a reference, the CGI application needs to specify the document's location with a URL. On receipt of the location header, the Web server will redirect the client to retrieve the document from the specified location. If the URL points to another document on the same server, then the server will retrieve the document from the specified location, and return the full document to the client.

Java servlet and EJB applications for database access

Java-based technologies have superseded CGI programming. The Java framework consists of Java servlets, Enterprise JavaBeans (EJBs), and the Java Database Connectivity (JDBC) API. A Java servlet is a Java class that can connect a Web server to other computational resources or to data stores. Java servlet-based applications are invoked by the Web server and have a life cycle that is managed by the servlet container, an execution environment that hosts the servlets and whose behavior is defined in the J2EE specification. Included in the execution environment are facilities for interfacing to a Web server. Standard J2EE Java packages provide the required support for implementing an application's interface to the Web server and to data servers.

Upon receiving a client request, the Web server facilities of the servlet container load, if not yet loaded, a servlet that can process the request. The servlet container will determine the particular servlet that gets loaded based on the URL in the client's request. A *service()* method in the servlet application gets called by the servlet container to process the client request. Depending on the particular client request (for example, an HTTP GET), a corresponding service method will be called from within *service()* (for example, *doget()* to process the HTTP GET). After extracting the request parameters, the servlet application issues the SQL queries via the standard JDBC API to retrieve or update data in a back-end database. Alternatively, the servlet can invoke an EJB to carry on the data access operation.

10.2.6 AAA server

According to the agreement between the Web service owner and the mobile network operator, the AAA functionality could be split between them and, as is often the case, it could be mostly under the operator's responsibility. Upon receiving a user initiated Web information request, the Application server sends a corresponding user authentication and authorization request to the AAA server. The AAA server could store a database of user names and passwords and validate a user based on the submitted credentials. A more automated way for authenticating a user would be to use certificate-based authentication. In this case, the mobile terminal sends the user's certificate and a digitally signed message to the Application server which, in turn, submits these items to the AAA server. This server extracts from the certificate the user's public key and uses it to decrypt the digitally signed message, hence authenticating the mobile user and ensuring data integrity.

After authenticating a user, the AAA server generates accounting service records that could include charges that are dependent on the amount of transferred data and the

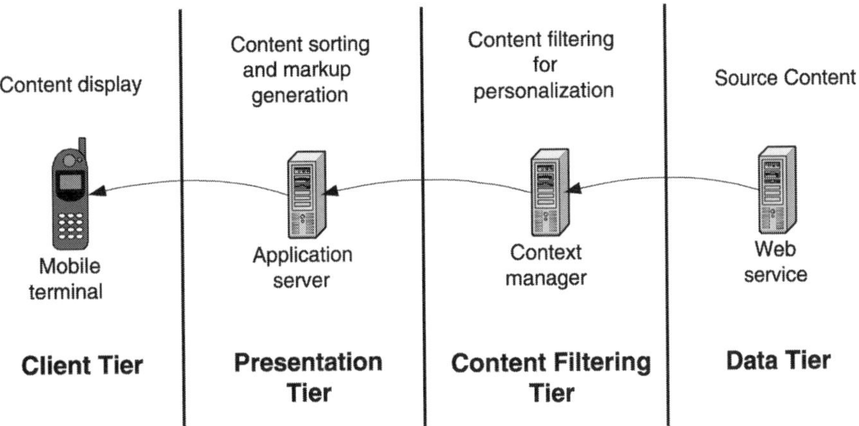

Figure 10.6 Information content data flow.

particular charging agreement that the user signed up to when subscribing to the Web service.

10.2.7 Wireless gateway

The mobile operator's network includes a wireless gateway that interfaces between the radio and infrastructure networks. In WAP 2.0, the HTTP and TCP protocols are used all the way between a mobile terminal and the content server. In this case, the wireless gateway provides for optimized wireless link communication by compressing the message body of HTTP responses.

The gateway also supports the establishment of a secure TLS tunnel between a mobile terminal and the content server when a secure connection is required. In addition, content push functionality can be provided by the gateway since it can host the PPG. The PPG controls access to mobile terminals from network content servers and resolves the addresses of mobile terminal recipients.

10.3 Message flow

The mobile Web services architecture partitions the network elements in a four-tiered system for data storage, context-based processing, presentation preparation, and display (Figure 10.6). The information data flow starts at the information source, in the Web Service, proceeds to the Context Manager for personalized filtering, and from there to the Application server where the information is transformed into a displayable format with a layout that is dependent on the particular content, and then ends up at the mobile

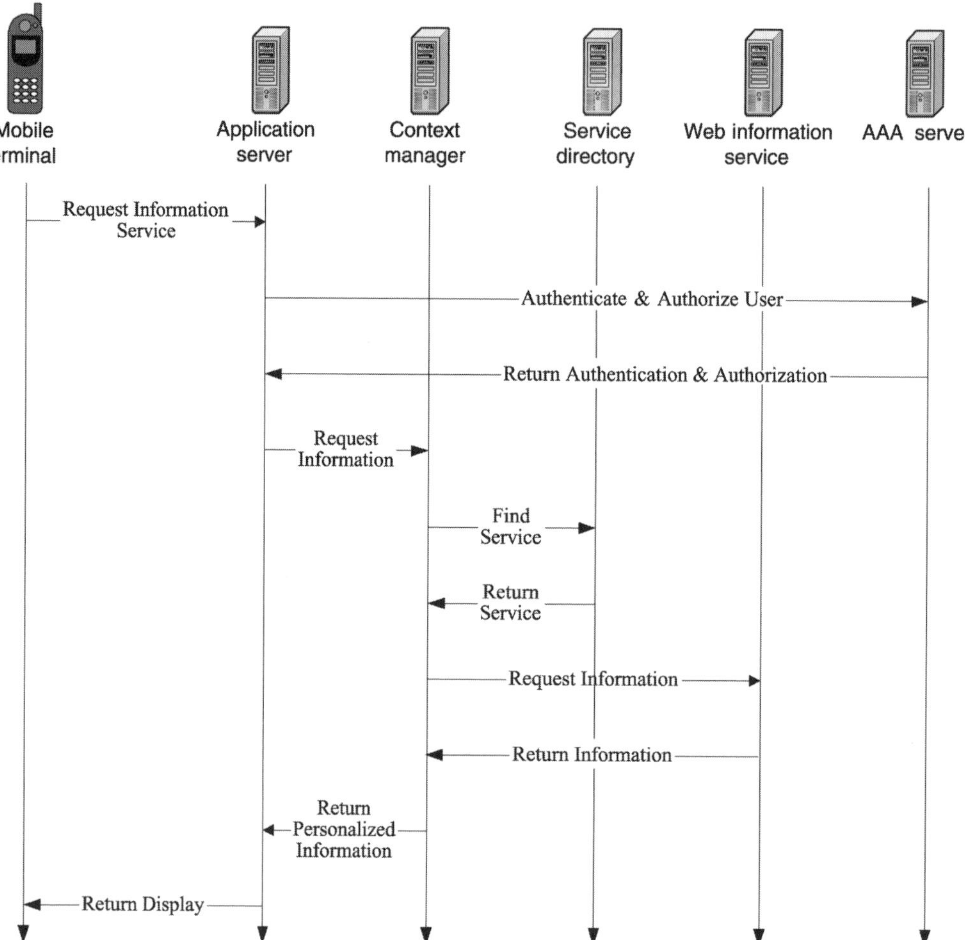

Figure 10.7 Web information request message flow.

terminal where it is displayed. Figure 10.6 shows this data flow and the corresponding role of each network element.

The message flows for an information request scenario in a mobile Web services architecture are shown in Figure 10.7. The mobile user starts by issuing a service request, for example to find local restaurants, to an Application server. The Application server, in turn, first authenticates the user by sending an authentication query to an AAA server. Upon receiving an approved authentication response, the Application server forwards the information request to the Context Manager where the user's preferences profile is stored.

The Context Manager finds an appropriate Web information service by querying a Service Directory; the selected information service could depend on user pre-specified preferences. After settling on a service, the Context Manager sends it an information request. The returned information is then filtered according to user preferences as listed in the user's profile. The personalized information is returned to the Application server for transformation into displayable markup. The Application server performs one last step of sorting the information so that the most relevant information items appear first. This last step is dependent upon the type of retrieved information. Different sorting criteria could be applied, for example, to restaurants and to hotels. Finally, the generated markup is sent to the mobile terminal for display.

Not shown in Figure 10.7 are the periodic context data updates that are sent from various sensors, or other devices, to the Context Manager. For example, if the mobile terminal is equipped with a GPS receiver, it can periodically send a location update to inform the Context Manager of its trajectory. A traffic server could similarly send to the Context Manager periodic updates of local traffic conditions so that affected users could adapt their travel plans correspondingly.

10.4 Network interfaces for seamless collaboration

Different efforts are underway in the Internet development community to facilitate the interoperability of applications. Many of today's mobile Web applications are in fact client/server based where a browser provides access to the data maintained by an Internet site. The Internet site will support stand-alone applications that provide all the services needed to generate the requested data.

The next generation of mobile Web applications will be composed of loosely coupled services that interoperate to deliver user requested information. This application model builds on a coarse grain of reusable components – a Web service. Examples of such services include an authentication service, and an encoding service; even a directory of services is a service too. Application developers will achieve much greater productivity when they reuse these services to build new applications. The protocols used to support mobile Web application communication are described in the following section.

10.4.1 SOAP inter-server protocol

Services need to expose interfaces that can be understood by a variety of applications, and data communication with these services needs to be done in a platform-independent way. XML technologies address these requirements since XML enables sharing of information between disparate services that are hosted on different platform types, and is therefore well suited for loosely coupled systems. XML remote procedure call (XML-RPC) is a specification for invoking a method on a remote server using XML [8]. Structured XML messages that encapsulate the remote method calls are sent over HTTP. An HTTP POST is used to convey the XML so that there is no problem with firewalls, as almost all firewalls will allow HTTP messages to pass through. The POST response message will convey back an XML representation of the method's response. Listings 10.1 and 10.2 show example call and response XML-RPC messages for retrieving a restaurant's cuisine type.

```
<?xml version="1.0"?>
<methodCall>
 <methodName>restaurants.getRestaurantCuisine</methodName>
 <params>
   <param>
     <value><string>By the Park</string></value>
   </param>
 </params>
</methodCall>
```

Listing 10.1 XML-RPC call message

```
<?xml version="1.0"?>
<methodResponse>
  <params>
    <param>
      <value><string>American</string></value>
    </param>
  </params>
</methodResponse>
```

Listing 10.2 XML- RPC response message

As can be seen, every value in XML-RPC is demarcated by a *<value>* element, and requires another element that indicates the value's type. Similarly, each part of a structure has a name and value pair.

The Simple Object Access Protocol (SOAP) protocol, defined by the W3C organization, specifies the format of standard XML messages used to communicate requests and responses among systems [9]. Because the message format is standardized and based on the XML standard, SOAP can be used to communicate among multiple computer architectures, languages, and operating systems. The SOAP specification enables a new class of applications, referred to as Web services, which expose services in a standard way so that application developers can create new applications by accessing these services on multiple Web resources. The CATIS network infrastructure components described in the next chapter are all implemented as Web services, and hence allow for remote invocation of their advertised services over the SOAP protocol.

The SOAP protocol has enhanced the XML-RPC proposal for remote invocation, however SOAP is not limited to just RPC-like communication and can be used to exchange XML documents. For example, Sun's Java API for XML messaging (JAXM) [10] implements SOAP messaging and can be used by a client to send an XML document to the server without specifying a remote method invocation.

SOAP defines the *<Envelope>* element, which is an XML construct that represents the contents of a SOAP message. The specification provides also an optional set of encoding rules that defines how language types are mapped to XML in a SOAP message, and an optional RPC format that defines how function calls are expressed in SOAP messages. With the availability of XML Schema [11], it is common practice to use literal XML Schema definitions that specify exactly how the request and response messages should be formatted in XML. Data typing as defined by XML Schema can convey more information about the transferred data as shown in the following string parameter:

```
<RestaurantName xsd:type="string">By the Park</RestaurantName>
```

SOAP also defines a protocol binding framework, which specifies how it can be bound onto another underlying protocol. The only defined binding allows the exchange of SOAP messages either as payload of a HTTP POST request and response, or as a SOAP message in the response to a HTTP GET. Almost all SOAP implementations include an HTTP binding, although the W3C specification doesn't prevent the use of other communication protocols to transport SOAP messages.

Listing 10.3 shows a simple SOAP request message that specifies a remote method and an associated parameter for retrieving a user's preferred restaurant cuisine, and Listing 10.4 shows the SOAP response message.

```
<?xml version="1.0" encoding="utf-8"?>
<env:Envelope
  xmlns:env="xmlns:env="http://www.w3.org/2003/05/
  soap-envelope"
  xmlns:xsi="http://www.w3.org/2001/XMLSchema-instance"
  xmlns:xsd="http://www.w3.org/2001/XMLSchema">

  <env:Header>
    <a:Account xmlns:a="some URI">
      <a:AccountGroup xsi:type="xsd:string">Chicago
      </a:AccountGroup>
    </a:Account>
  </env:Header>

  <env:Body>
    <b:GetCustomerPreferedCuisine xmlns:b=
      "some other URI">
```

```
      <b:CustomerID xsi:type="xsd:int">555-123-4567
      </b:CustomerID>
    </b:GetCustomerPreferedCuisine>
  </env:Body>
</env:Envelope>
```

Listing 10.3 SOAP request message

```
<?xml version="1.0" encoding="utf-8"?>
<env:Envelope
  xmlns:env="xmlns:env="http://www.w3.org/2003/05/
  soap-envelope"
  xmlns:xsi="http://www.w3.org/2001/XMLSchema-instance"
  xmlns:xsd="http://www.w3.org/2001/XMLSchema">

  <env:Body>
    <b:GetCustomerPreferedCuisineResponse xmlns:b="some
     other URI">
      <b:CustomerCuisine xsi:type="xsd:string">Italian
      </b:CustomerCuisine>
    </b:GetCustomerPreferedCuisineResponse>
  </env:Body>
</env:Envelope>
```

Listing 10.4 SOAP response message

Listing 10.3 shows that the SOAP envelope contains both a *<Header>* and *<Body>* sub-elements. The *<Header>* element, which is optional, can hold control elements for handling login, security, transaction, and billing concerns, while the *<Body>* element contains the application's message. The header can specify whether intermediate network nodes or the receiver should deal with parts of the message. Intermediate nodes can process the header according to a control function they are assuming, for example, a billing function, while the receiver node processes the body.

In the provided example, the receiver uses the header data to process accounting information related to the request. The remote method name is the name of the request element *<GetCustomerPreferredCuisine>* and the associated parameter value is the *<CustomerID>* element's value. The *<Body>* element in the SOAP response of Listing 10.4 provides the user's preferred cuisine in an element that has *Response* appended to the name of the request element from the request message.

SOAP typically uses HTTP as the transport mechanism, and the SOAP message is the body of a POST command with the content type *text/xml* or *application/soap+xml* (in the latest SOAP 1.2). No special ports need to be configured on the server side, and most firewalls that allow HTTP through will let SOAP messages through, unless filtering is done on the SOAP messages. The HTTP message includes a mandatory header that must be understood by the receiving server when processing the SOAP request. This header is the *SOAPAction* HTTP header, and it specifies the Web service that is invoked on the server side.

10.4.2 WSDL service description

In order to successfully call a Web service a client needs to know the service address, what operations the service supports, what parameters the service expects, and what the service returns. The Web Service Definition Language (WSDL), defined by the W3C organization, provides all of this information in an XML document that can be read by humans or can be machine-processed [12]. The WSDL specification includes two parts. The first part describes the specification of interactions with the service in the form of sets of messages exchanged between the client and the service. The XML service description can be bound to a concrete network protocol and message format to define a service endpoint. Supported protocols include SOAP, HTTP, and SMTP. The second part of the specification describes sequencing and cardinality of the messages involved in a particular interaction.

A WSDL document describes Web services starting with the messages that are exchanged between the service provider and the client. An exchange of messages between the service provider and the client is described as an operation, and a collection of operations is called an interface. Listing 10.5 shows the WSDL 1.2 [13] XML elements of an interface describing a hotel finder service. In the latest WSDL 2.0 [12], some elements were renamed, but otherwise the WSDL structure is the same.

```
<?xml version="1.0" encoding="UTF-8"?>
<definitions name="LodgingFinder"
  targetNamespace="http://LodgingFinder.org/wsdl"
  xmlns:tns="http://LodgingFinder.org/wsdl"
  xmlns:xsd="http://www.w3.org/2001/XMLSchema"
  xmlns:soap="http://schemas.xmlsoap.org/wsdl/soap/"
  xmlns="http://schemas.xmlsoap.org/wsdl/">

  <message name="FindHotel">
    <part name="city" type="xsd:string"/>
    <part name="persons" type="xsd:int"/>
    <part name="rateLimit" type="xsd:int"/>
  </message>

```

```
  <message name="FindHotelResponse">
    <part name="result" type="xsd:string"/>
  </message>

  <portType name="FindLodging">
    <operation name="Hotel" pattern="http://www.w3.org/
      2003/11/wsdl/in-out">
      <input message="tns:FindHotel"/>
      <output message="tns:FindHotelResponse"/>
    </operation>
  </portType>

</definitions>
```

Listing 10.5 WSDL of a hotel finder Web service

The WSDL of Listing 10.5 defines two messages. The first *<message>* element is a request to find a hotel in a given city that can accommodate a specified number of persons below a certain room rate, and the second *<message>* element is the response containing the hotel name. The *<portType>* definition follows with an *<operation>* element that specifies the *<input>* and *<output>* messages, as well as the message sequence in the *pattern* attribute.

In addition to describing message contents, WSDL defines, via *<service>* elements, where the service is available and what communication protocols are used to talk to the service. The *<service>* element contains a collection of *<port>* elements, where each port connection accepts messages containing either document information or procedure-call information. Each *<port>* element refers to a *<binding>* element that binds an operation to a concrete protocol and message format. The *<port>* element also contains the URI address of where the operation can be invoked. Since a service may contain multiple ports, it can expose its services through multiple protocols (SOAP, HTTP, and SMTP). Listing 10.6 shows a simplified description of a service with one port that can process the previously specified SOAP messages for finding hotels, and is accessed at a URI specified by the *<soap:address>* element.

```
<definitions>
  <binding name="LodgeBinding" interface=
    "FindLodging"type="tns:FindLodging">
  <soap:binding transport="http://schemas.xmlsoap.org/
    soap/http"style="rpc"/>
  <operation name="Hotel">
    <soap:operation soapAction=
      "http://LodgingFinder.org/HotelFinder"/>
```

```
      <input>
         <soap:body encodingStyle="http://schemas.xmlsoap.
          org/soap/encoding/" use="encoded"
          namespace="http://LodgingFinder.org/wsdl"/>
       </input>
       <output>
         <soap:body encodingStyle="http://schemas.xmlsoap.
          org/soap/encoding/" use="encoded"
          namespace="http://LodgingFinder.org/wsdl"/>
       </output>
     </operation>
   </binding>
   <service name="LodgeEndpoint">
     <port name="LodgeEndpoint"
      binding="tns:LodgeBinding">
       <soap:address location="http://LodgingFinder.org/
        HotelFinder"/>
     </port>
   </service>
   </definitions>
```

Listing 10.6 WSDL for binding and service elements

A number of tools were developed to enable the generation of client and server skeleton code from WSDL descriptions. For example, Sun's Java API for XML RPC (JAX-RPC) [10] implementation includes a *wscompile* tool that uses a WSDL document to generate *stubs*, the template classes that are needed by a client to communicate with a remote service. The JAX-RPC implementation has another tool, called *wsdeploy*, that creates *ties*, the template classes that the server needs to communicate with a remote client.

The JAX-RPC runtime system uses the stubs and ties created by *wscompile* and *wsdeploy* behind the scenes. First, it converts, on the client side, the client's remote method call into a SOAP message and sends it to the Web service as an HTTP request. On the server side, the JAX-RPC runtime system receives the request, translates the SOAP message into a method call, and invokes it. After the Web service has processed the request, the runtime system goes through a similar set of steps to return the result to the client. As complex as the implementation details of communication between the client and server may be, they are invisible to both the Web services and their clients.

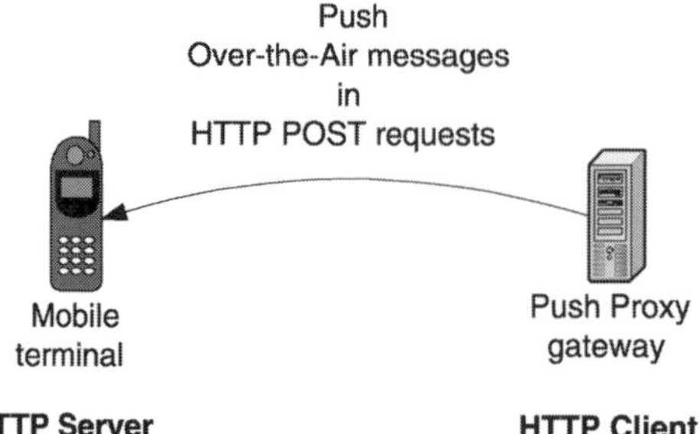

Figure 10.8 Mobile terminal as HTTP server of push messages.

10.4.3 Client–server protocols

The mobile terminal accesses network server applications either from its Web browser or from a stand-alone application running on the terminal. If access is through the Web browser, then the HTTP protocol is used for retrieving multimedia content or returning form content.

When a stand-alone application is executing on the mobile terminal, it can access remote application operations via the SOAP protocol in the same way that inter-server calls are being processed. The benefits of using SOAP on the client are the same as for the server case: there is no dependency on the remote server operating system or programming language. An example software platform that supports SOAP on wireless clients is Sun's Java 2 Platform Micro-Edition (J2ME) version of Java for mobile terminals, as J2ME was extended with support for Web services. With this extension, J2ME clients can issue SOAP messages to network servers.

If information transfer requests are initiated in the network, then the Push Over-the-Air Protocol [5] is used to send content to the client (Figure 10.8). These push messages are encapsulated by the Push Proxy Gateway (PPG) server in HTTP POST requests, so that the server acts as an HTTP client and the mobile terminal as an HTTP server. To enable push delivery, an active TCP connection needs to be established between the mobile terminal and the PPG. There are two ways to establish such a connection; a PPG originated TCP or a mobile terminal originated TCP. In the terminal-originated approach, a request is sent to the terminal, for example, in the form of a short message service (SMS) message, to set up the connection.

REFERENCES AND FURTHER READING

[1] Sun Microsystems Inc., *Sun Java System Application Server*, http://www.sun.com/software/products/appsrvr/home_appsrvr.html.

[2] WAP Forum, *WAP Push Architectural Overview*. WAP-250-PushArchOverview-20010703-a. (http://www.openmobilealliance.org/tech/affiliates/wap/wapindex.html, Jul. 3, 2001).

[3] WAP Forum, *Push Access Protocol*. WAP-247-PAP-20010429-a. (http://www.openmobilealliance.org/tech/affiliates/wap/wapindex.html, Apr. 29, 2001).

[4] WAP Forum, *Push Proxy Gateway Service*. WAP-249-PPGService-20010713-a. (http://www.openmobilealliance.org/tech/affiliates/wap/wapindex.html, Jul. 13, 2001).

[5] WAP Forum, *Push OTA Protocol*. WAP-235-PushOTA-20010425-a. (http://www.openmobilealliance.org/tech/affiliates/wap/wapindex.html, Apr. 25, 2001).

[6] OASIS, *UDDI*, http://uddi.org/about.html.

[7] OASIS, *UDDI Version 3.0.1 Committee Specification*, http://www.oasis-open.org/committees/uddi-spec/doc/tcspecs.htm#uddiv3 (Oct. 14, 2003).

[8] UserLand Software Inc., *XML-RPC Specification*, http://www.xmlrpc.com/spec.

[9] M. Gudgin *et al., SOAP Version 1.2 Part 1: Messaging Framework*. W3C Recommendation (Jun. 24, 2003), http://www.w3.org/TR/soap12-part1/.

[10] Sun Microsystems Inc., *Reference APIs*, http://developers.sun.com/techtopics/webservices/reference/api/index.html.

[11] P. Biron *et al., XML Schema Part 2: Datatypes*. W3C Recommendation (May 2, 2001), http://www.w3.org/TR/xmlschema-2/.

[12] R. Chinnici *et al., Web Services Description Language (WSDL) Version 2.0 Part 1: Core Language*. W3C Working Draft (Nov. 10, 2003), http://www.w3.org/TR/wsdl20.

[13] R. Chinnici *et al., Web Services Description Language (WSDL) Version 1.2*. W3C Working Draft (Mar. 3, 2003), http://www.w3.org/TR/2003/WD-wsdl12-20030303.

Sun Microsystems Inc., *Technical Topics Web Services*, http://developers.sun.com/techtopics/webservices/index.html.

J. Moreau and J. Schlimmer, *Web Services Description Language (WSDL) Version 1.2 Part 3: Bindings*. W3C Working Draft (Jun. 11, 2003), http://www.w3.org/TR/wsdl12-bindings.

11 Context-aware tourist information system

In this chapter, we elaborate on the CATIS experimental mobile Web network. CATIS implements a context-aware architecture that delivers tourism information to mobile users in a city environment and leverages context data to adapt the delivered Web information. The user scenarios that CATIS caters for are of tourists who search for nearby restaurants which meet their pre-specified preferences. The CATIS components include mobile terminals, an application server, a context and profile manager, a service directory, and Web information services. In the following sections, we describe the functionality of these network elements and the message flows between them. We then proceed with a description of the system interfaces and the content managed by each of these elements.

11.1 User scenarios

The Context-Aware Tourist Information Service (CATIS) is an experimental mobile Web network ([1], [2], and [3]) developed at Northwestern University. The scenarios implemented in CATIS consist of mobile tourists that request restaurant information in their vicinity. The system adapts the displayed information to the tourist's personal preferences and speed of travel, as the tourist may be driving, walking, or using public transportation. The first screen presented on the tourist's mobile terminal displays a list of restaurants in the detected direction of travel. Figure 11.1 shows a PDA-type terminal screen with an ordered list of restaurants in the tourist's vicinity that meet his or her stated preferences. Selection of a particular restaurant from the list will bring up a new screen with the restaurant's detailed information.

Figure 11.2 shows detailed information of the first restaurant from the previous list. Information such as wait time for seating could be displayed on this screen, provided such information is available.

Figure 11.1 Recommended restaurants in the user's vicinity.

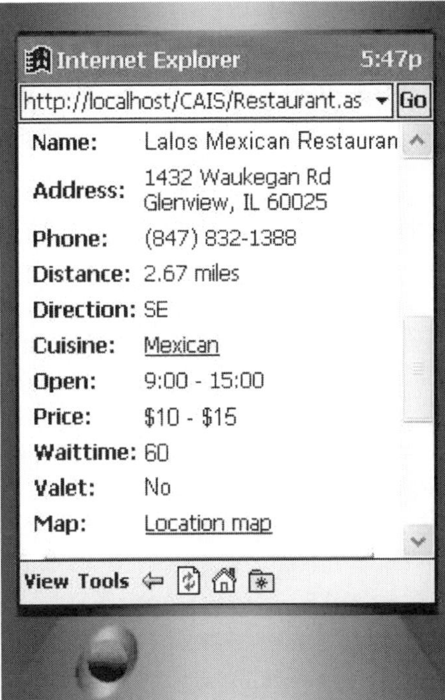

Figure 11.2 Detailed restaurant information.

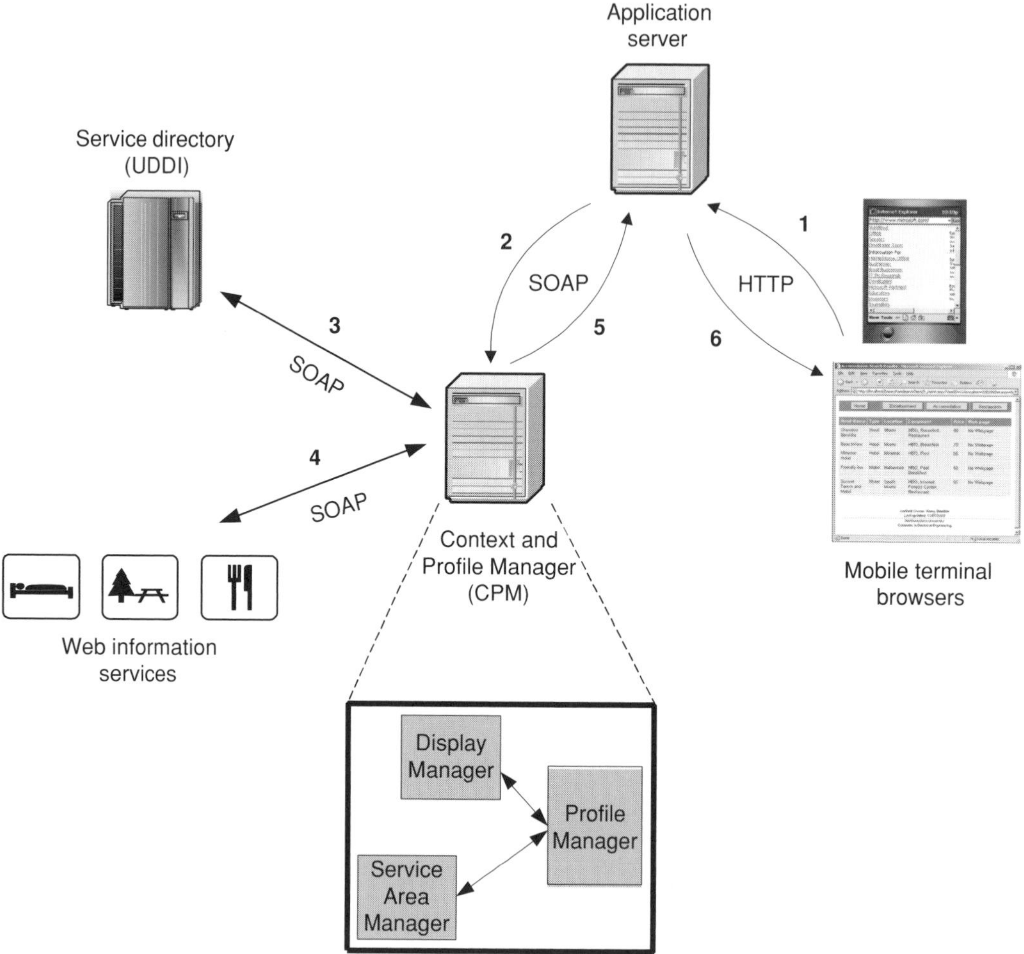

Figure 11.3 The CATIS architecture.

11.2 Architecture overview

The CATIS mobile network consists of five major components (Figure 11.3). The network infrastructure components are all implemented as Web services, and hence allow for remote invocation of their advertised services over the SOAP protocol. Access to and from the user's mobile terminal is different as it communicates with the infrastructure network elements with the HTTP protocol's GET and POST requests.

As advocated in the previous chapter, different functions are allocated to different network servers. In particular, as shown in the following sections, context processing and browser display generation are handled on distinct servers. This allocation promotes application development flexibility and network scalability.

11.2.1 Network elements

Mobile terminals

The mobile terminals are thin client browser-based devices. The terminals are GPS-enabled so that they can provide location data, and hence allow for the retrieval of content that is pertinent to their location. The mobile terminals issue periodic location updates to the Context and Profile Manager (CPM) where their trajectories are stored.

Application server

The Application server is the front-end of CATIS. This server registers new users, collects their preferences in Web forms and forwards this information to the CPM for storage in user profiles. Registered users make information requests addressed to the Application server, which are then sent to the CPM for further processing. Upon receiving back the user requested information, this server prepares the information for display by sorting the list of items according to user preferences, and generating the markup for the user's mobile terminal browser.

Context and Profile Manager (CPM)

The hub of the mobile network is a Context and Profile Manager (CPM). The CPM is responsible for the back-end information processing: it generates the context-filtered Web-based information that is sent to mobile users. The main functional components of the CPM are shown in the enlarged CPM components box of Figure 11.3. These components include a Service Area Manager, a Profile Manager, and a Display Manager.

To personalize the delivered information, the Profile Manager maintains a database of user profiles. The profiles contain both static preferences and dynamic context. The Profile Manager collects user preferences at user registration time, and dynamic context, such as location, at run time. All context data is stored in the respective user profiles. The Service Area Manager receives from the Application server user requests for information in the mobile user's vicinity, and finds for each request a suitable Web information service by querying a service directory. It then determines the user's relevant geographic service area, and asks a selected Web information service for the location-based information content. The retrieved content is then filtered by the Profile Manager by the user interests as expressed in his or her profile, and forwarded for display to the Application server. Finally, the Display Manager is responsible for periodic updates of the user's terminal screen.

Service directory

The Service directory enables the registration and discovery of Web information services. It is implemented as a UDDI server. A client application can search for certain businesses or services in the UDDI directory through a Web service interface. The returned address of a found business is a URL where the business can be contacted for service requests.

Web information services

The Web information services provide the content that the mobile user is seeking. Included are a restaurant finder, a hotel finder, and an entertainment park locator. Since these services are all very alike from a Web service implementation perspective, we describe herein just the restaurant finder service.

11.2.2 Message data flows

The following message sequence corresponds to the numbering in Figure 11.3 and reflects a typical scenario of a mobile tourist that requests to find a restaurant in his vicinity. The message sequence assumes that the mobile user has previously registered preferences with CATIS.

1. The mobile user logs on to the Application server and issues a query for restaurants at a current location.
2. The Application server validates the user and forwards the request to the CPM's Service Area Manager.
3. The CPM's Service Area Manager issues a request to the Service directory to get a list of registered Web information services that can find restaurants. Since the requested information can be provided in multiple formats, for example, text or audio, to fit the changing user situation, the Service Area Manager selects the appropriate service.
4. The CPM's Service Area Manager selects a restaurant Web information service and issues a query for restaurants in the vicinity of the user. The area from which restaurants are selected is referred to as the user's personal service area.
5. The retrieved Web content is filtered by the user's interests by the CPM's Profile Manager and augmented with relevant context information such as distance from user. The CPM's Profile Manager then returns the filtered content to the Application server.
6. The Application server sorts the list of restaurants by user preferences, so that top choices appear first on the list. It then generates markup appropriate for the user's terminal browser, and sends it to the terminal.

At the completion of the last message sequence, the mobile user scans the returned list and picks a restaurant of his or her liking.

11.3 System interfaces

All the network elements, except for the mobile terminal, are implemented as Web services. As such, they have a programmatic interface that allows calling their offered services via remote procedure calls. A WSDL file is also associated with each network element interface. The file is generated from a corresponding Java interface, that provides all the pertinent information about a Web service. This information includes the Web service name, the operations that can be called on it, the parameters for those operations, and the URI of where to send requests. Clients use the WSDL of a server to generate the corresponding code stubs that contain the classes needed to communicate with the server. The WSDL is also used on the server side to generate the peer code ties that contain classes needed to receive the client requests. The CATIS Java interfaces and associated WSDL files are described in the following sections.

11.3.1 Restaurant Web service interface

The restaurant service provides a *Search* call that returns all the restaurants that are contained in a rectangular area specified by the latitude and longitude of its sides. The restaurant Web service interface is shown in Listing 11.1.

```
public interface RestaurantWebService extends Remote
{
  public String Search ( double latitudeMin, double
    latitudeMax, double longitudeMin, double longitudeMax,
    String time, String[] cuisines)
    throws RemoteException;
}
```

Listing 11.1 Restaurant Web service Java interface

The *time* and *cuisines* parameters are optional and are used only for restaurant services that can filter the search according to the restaurants' opening hours and offered cuisines.

The corresponding service WSDL file is shown in Listing 11.2.

```
<?xml version="1.0" encoding="UTF-8"?>
<definitions xmlns="http://schemas.xmlsoap.org/wsdl/"
  xmlns:tns="http://www.awarenetworks.com/wsdl/RestaurantWebService"
  xmlns:ns2="http://www.awarenetworks.com/xsd/RestaurantWebService"
  xmlns:xsd="http://www.w3.org/2001/XMLSchema"
  xmlns:soap="http://schemas.xmlsoap.org/wsdl/soap/"
   name="RestaurantWebService"
  targetNamespace="http://www.awarenetworks.com/wsdl/
   RestaurantWebService">
  <types>
    <schema xmlns="http://www.w3.org/2001/XMLSchema"
      xmlns:tns="http://www.awarenetworks.com/xsd/RestaurantWebService"
      xmlns:soap11-enc="http://schemas.xmlsoap.org/soap/encoding/"
      xmlns:xsi="http://www.w3.org/2001/XMLSchema-instance"
      xmlns:wsdl="http://schemas.xmlsoap.org/wsdl/"
```

```
    targetNamespace="http://www.awarenetworks.com/xsd/
      RestaurantWebService">
<import namespace="http://schemas.xmlsoap.org/soap/encoding/"/>
<complexType name="ArrayOfstring">
  <complexContent>
    <restriction base="soap11-enc:Array">
      <attribute ref="soap11-enc:arrayType"
        wsdl:arrayType="string[]"/>
    </restriction>
  </complexContent>
</complexType>
</schema>
</types>

<message name="RestaurantWebService_Search">
  <part name="latitudeMin" type="xsd:double"/>
  <part name="latitudeMax" type="xsd:double"/>
  <part name="longitudeMin" type="xsd:double"/>
  <part name="longitudeMax" type="xsd:double"/>
  <part name="time" type="xsd:string"/>
  <part name="cuisines" type="ns2:ArrayOfstring"/>
</message>

<portType name="RestaurantWebService">
  <operation name="Search"
    parameterOrder="latitudeMin latitudeMax longitudeMin
      longitudeMax time
                    cuisines">
  <input message="tns:RestaurantWebService_Search"/>
  <output message="tns:RestaurantWebService_SearchResponse"/>
  </operation>
</portType>

<binding name="RestaurantWebServiceBinding"
  type="tns:RestaurantWebService">
  <operation name="Search">
  <input>
    <soap:body encodingStyle="http://schemas.xmlsoap.org/
      soap/encoding/"
        use="encoded"
        namespace="http://www.awarenetworks.com/wsdl/
          RestaurantWebService"/>
  </input>
  <output>
```

```
      <soap:body encodingStyle="http://schemas.xmlsoap.org/
        soap/encoding/"
          use="encoded"
          namespace="http://www.awarenetworks.com/wsdl/
            RestaurantWebService"/>
    </output>
    <soap:operation soapAction=""/>
  </operation>
  <soap:binding transport="http://schemas.xmlsoap.org/soap/http"
    style="rpc"/>
  </binding>

  <service name="RestaurantWebService">
    <port name="RestaurantWebServicePort"
      binding="tns:RestaurantWebServiceBinding">
      <soap:address xmlns:wsdl="http://schemas.xmlsoap.org/wsdl/"
        location="http://localhost:8980/RestaurantWebService/
          RestaurantWebService"/>
    </port>
  </service>
</definitions>
```

Listing 11.2 Restaurant Web service WSDL

11.3.2 Application server interface

The Application server provides a *notify* call to inform it that it needs to update the displayed list of restaurants on the user's terminal. The call is performed by the CPM's Display Manager that tracks the user's speed of movement, and adjusts correspondingly the rate at which updates should be performed. For faster moving users, the update rate can be as short as once per minute, so that a displayed list adapts appropriately to the new areas traversed by the user. The Application server interface is shown in Listing 11.3.

```
public interface AppServer extends Remote {
  public void notify( String username ) throws RemoteException;
}
```

Listing 11.3 Application server Java interface

The *username* parameter identifies the particular user for which the display should be updated.

The corresponding service WSDL file is shown in Listing 11.4.

```
<?xml version="1.0" encoding="UTF-8"?>
<definitions xmlns="http://schemas.xmlsoap.org/wsdl/"
  xmlns:tns="http://www.awarenetworks.com/wsdl/
    adaptationServer"
  xmlns:xsd="http://www.w3.org/2001/XMLSchema"
  xmlns:soap="http://schemas.xmlsoap.org/wsdl/soap/"
    name="adaptationServer"
  targetNamespace="http://www.awarenetworks.com/wsdl/
    adaptationServer">
  <types/>

  <message name="AppServer_notify">
    <part name="username" type="xsd:string"/>
  </message>
  <message name="AppServer_notifyResponse"/>

  <portType name="AppServer">
    <operation name="notify" parameterOrder="username">
      <input message="tns:AppServer_notify"/>
      <output message="tns:AppServer_notifyResponse"/>
    </operation>
  </portType>

  <binding name="AppServerBinding" type="tns:AppServer">
    <operation name="notify">
      <input>
        <soap:body encodingStyle="http://schemas.xmlsoap.org/
          soap/encoding/"
            use="encoded"
            namespace="http://www.awarenetworks.com/wsdl/
              adaptationServer"/>
      </input>
      <output>
        <soap:body encodingStyle="http://schemas.xmlsoap.org/
          soap/encoding/"
            use="encoded"
            namespace="http://www.awarenetworks.com/wsdl/
              adaptationServer"/>
      </output>
      <soap:operation soapAction=""/>
    </operation>
    <soap:binding transport="http://schemas.xmlsoap.org/soap/http"
      style="rpc"/>
  </binding>

  <service name="AdaptationServer">
```

```
    <port name="AppServerPort"binding="tns:AppServerBinding">
      <soap:address xmlns:wsdl="http://schemas.xmlsoap.org/wsdl/"
        location="http://localhost:8980/adaptationServer/
          adaptationServer"/>
    </port>
  </service>
</definitions>
```

Listing 11.4 Application server WSDL

11.3.3 Context and profile manager interface

The CPM provides an interface where the majority of the calls are for registering a new user, updating information in a profile, retrieving information from a profile, and deregistering a user. Other calls request Web content information, or notifications about the availability of new Web content. All calls to the CPM are issued by the Application server. Example calls that are included in the CPM's interface are shown in Listing 11.5.

```
public interface CPM extends Remote
{
  public boolean addAddresses ( String username, String password,
    Address[] addrs )
    throws RemoteException;
  public Address[] getAddresses ( String username, String password )
    throws RemoteException;

  public boolean insertNewLocation ( String username, String password,
    double gpsLatitude, double gpsLongitude )
    throws RemoteException;
  public boolean updateLocalTime(String username, String password,
    java.util.Calendar timestamp)
    throws RemoteException;

  public String getWebServiceInformation(String username, String password,
    String webServiceType)
    throws RemoteException;

  public boolean registerForEvent(String username, String password,
    String[] params)
    throws RemoteException;

  //MANY MORE METHODS...
}
```

Listing 11.5 Context and Profile Manager Java interface

The *addAddresses* and *getAddresses* calls are for updating the user's profile with new address information, and for retrieving address information, respectively. Example postings of dynamic context data are the *insertNewLocation* call that updates the user's profile with the user's geographical coordinates and the *updateLocalTime* call that updates the profile with the user's current time stamp. The last two calls are issued in sequence so that the user's speed of movement can be estimated from successive updates.

The *getWebServiceInformation* call is issued by the Application server after the user has requested location-based information. The type of requested data, for example, *restaurant*, is specified in the *webServiceType* parameter. Finally, the CPM provides the *registerForEvent* call to clients that request to be informed of specific event occurrences. For example, the Application server wants to be notified when the list of restaurants on the user's display should be updated. The CPM tracks the mobile user's travel, and uses predefined criteria that will trigger such notifications. For example, the frequency of updates could be directly related to the user's speed.

The user's profile contains attributes that describe a user's preferences and dynamic context information. The profile's Java class structure is shown in Listing 11.6.

```
public class Profile extends XMLStructure
{
   //PRIVATE ATTRIBUTES
   private String description="";
   private UserInfo userInfo = new UserInfo();
   private Address[] addresses;
   private PhoneNumber[] phoneNumbers;
   private Email[] emailAddresses;
   private Preference[] preferences;
   private Location[] locationHistory;
   private double speed = 0;
   private String speedUnit = "mph";
   private double direction = 0;
   private String directionUnit = "dec-deg";
   private java.util.Calendar localTime = null;
   private String actualState = "";
   private String userDevice = "";

   //PUBLIC METHODS
   public void setUserInfo(UserInfo ui);
   public UserInfo getUserInfo();
   public boolean addAddress(Address addr);
   public void setAddresses(Address[] addrs);
   public Address[] getAddresses();
```

```
   //...many more getter and setter methods to get and set the
   //...private attributes
}
```

Listing 11.6 Context and Profile Manager Java class user profile

A user's address attributes are defined in a separate *Address* class. This class is shown in Listing 11.7.

```
public class address extends XMLStructure
{
   //PRIVATE ATTRIBUTES
   private String type = "";
   private String street = "";
   private String aptNo = "";
   private String postalCode = "";
   private String city = "";
   private String region = "";
   private String country = "";
   //PUBLIC METHODS

   public void setStreet( String street );
   public String getStreet();
   public void setAptNo( String aptNo );
   public String getAptNo();
   public void setPostalCode( String postalCode );
   public String getPostalCode();

   //...many more getter and setter methods to get and set the
   //...private attributes
}
```

Listing 11.7 Context and Profile Manager address class

The corresponding CPM WSDL file is shown in Listing 11.8.

```
<?xml version="1.0" encoding="UTF-8"?>
<definitions xmlns="http://schemas.xmlsoap.org/wsdl/"
    xmlns:tns="http://www.awarenetworks.com/wsdl/CPM"
    xmlns:soap="http://schemas.xmlsoap.org/wsdl/soap/"
    xmlns:xsd="http://www.w3.org/2001/XMLSchema"
    xmlns:ns2="http://www.awarenetworks.com/xsd/CPM" name="CPM"
    targetNamespace="http://www.awarenetworks.com/wsdl/CPM">
<types>
  <schema xmlns="http://www.w3.org/2001/XMLSchema"
```

```
    xmlns:tns="http://www.awarenetworks.com/xsd/CPM"
    xmlns:xsi="http://www.w3.org/2001/XMLSchema-instance"
    xmlns:soap-enc="http://schemas.xmlsoap.org/soap/encoding/"
    xmlns:wsdl="http://schemas.xmlsoap.org/wsdl/"
        targetNamespace="http://www.awarenetworks.com/xsd/CPM">

    <complexType name="Address">
      <sequence>
        <element name="aptNo" type="string"/>
        <element name="city" type="string"/>
        <element name="country" type="string"/>
        <element name="postalCode" type="string"/>
        <element name="region" type="string"/>
        <element name="street" type="string"/>
        <element name="type" type="string"/>
      </sequence>
    </complexType>

    <complexType name="ArrayOfAddress">
      <complexContent>
        <restriction base="soap-enc:Array">
          <attribute ref="soap-enc:arrayType"
            wsdl:arrayType="tns:Address[]"/>
        </restriction>
      </complexContent>
    </complexType>

    <!-- MANY MORE COMPLEX TYPES --!>
  </schema>
</types>

<message name="CPM_addAddresses">
  <part name="String_1" type="xsd:string"/>
  <part name="String_2" type="xsd:string"/>
  <part name="arrayOfAddress_3" type="ns2:ArrayOfAddress"/>
</message>
<message name="CPM_addAddressesResponse">
    <part name="result" type="xsd:boolean"/>
</message>

<message name="CPM_getAddresses">
  <part name="String_1" type="xsd:string"/>
  <part name="String_2" type="xsd:string"/>
</message>
<message name="CPM_getAddressesResponse">
  <part name="result" type="ns2:ArrayOfAddress"/>
</message>

<!-- MANY MORE message ELEMENTS --!>
```

```
<portType name="CPM">
  <operation name="addAddresses" parameterOrder="String_1 String_2
    arrayOfAddress_3">
    <input message="tns:CPM_addAddresses"/>
    <output message="tns:CPM_addAddressesResponse"/>
  </operation>
  <operation name="getAddresses" parameterOrder="String_1 String_2">
    <input message="tns:CPM_getAddresses"/>
    <output message="tns:CPM_getAddressesResponse"/>
  </operation>
</portType>
<!-- MANY MORE operation ELEMENTS --!>
<binding name="CPMBinding" type="tns:CPM">
  <operation name="addAddresses">
    <input>
      <soap:body encodingStyle="http://schemas.xmlsoap.org/soap/encoding/"
        use="encoded" namespace="http://www.awarenetworks.com/wsdl/CPM"/>
    </input>
    <output>
      <soap:body encodingStyle="http://schemas.xmlsoap.org/soap/encoding/"
        use="encoded" namespace="http://www.awarenetworks.com/wsdl/CPM"/>
    </output>
    <soap:operation soapAction=""/>
    </operation>
    <operation name="getAddresses">
    <input>
      <soap:body encodingStyle="http://schemas.xmlsoap.org/soap/encoding/"
        use="encoded" namespace="http://www.awarenetworks.com/wsdl/CPM"/>
    </input>
    <output>
      <soap:body encodingStyle="http://schemas.xmlsoap.org/soap/encoding/"
        use="encoded" namespace="http://www.awarenetworks.com/wsdl/CPM"/>
    </output>
    <soap:operation soapAction=""/>
    </operation>
  </binding>
<!-- MANY MORE binding ELEMENTS --!>
<service name="CPM">
  <port name="CPMPort" binding="tns:CPMBinding">
    <soap:address xmlns:wsdl="http://schemas.xmlsoap.org/wsdl/"
      location="http://localhost:8980/CPM/CPM"/>
  </port>
  </service>
</definitions>
```

Listing 11.8 Context and Profile Manager WSDL

11.4 Network elements: content and procedures

11.4.1 Restaurant Web service content

The restaurant Web service uses an Apache Xindice database [4] for storing its content data. Xindice is a native XML database designed to store XML data directly. No schema needs to be defined, and the XML files are stored in database collections; access is available through a Java API defined by XML:DB [5]. For the prototype, data was collected from the greater Chicago area, and included a sample of 2200 restaurants. Readily available restaurant attribute includes information about:

- restaurant name, address, and phone number;
- cuisine;
- price range.

This information was augmented with the following attributes, when available:

- restaurant GPS coordinates (latitude and longitude) derived from the address;
- a map showing the restaurant's location;
- a restaurant rating factor;
- opening hours;
- whether valet parking is offered.

An example XML structure of a restaurant's information is shown in Listing 11.9.

```
<Restaurant>
  <Name>Tanellis Chicago Style Pizza</Name>
  <Street>5807 W Diversey Ave</Street>
  <City>Chicago</City>
  <State>IL</State>
  <Zip>60639</Zip>
  <Phone>(773) 622-4244</Phone>
  <Cuisine>Pizza</Cuisine>
  <Latitude>41.931063</Latitude>
  <Longitude>-87.771360</Longitude>
  <YahooMap>http://imgi.maps.yahoo.com/mapimage?MAPData=ccjS
    Uvhyzy286JaOg23zy23YX1j.Mou6fry7xGSzvdH4a5X3yemK5XBI0j2oJ
    q37aqcygoVqwG_oxBs6dSdG51NsqQOa_Klbjeu21a4z7lwIXRGEufOFu.
    jZTJclveox</YahooMap>
  <Rating>
    <Overall>-</Overall>
    <Food>-</Food>
    <Service>-</Service>
    <Ambience>-</Ambience>
  </Rating>
  <PriceLow>$10</PriceLow>
```

```
    <PriceHigh>$18</PriceHigh>
    <Open>9:00</Open>
    <Close>22:30</Close>
    <Valet>0</Valet>
</Restaurant>
```

Listing 11.9 Restaurant data in XML

11.4.2 Context and profile manager content

Like the restaurant Web service, the CPM stores its content in an Apache Xindice database [4]. The CPM's content consists of user profiles in the form of XML structures. Listing 11.10 shows an example user profile.

```
<Profile>
  <Description>Profile for test user</Description>
  <Static>
    <UserName>superman</UserName>
    <Password>1234</Password>
    <LastName>Kent</LastName>
    <FirstName>Clark</FirstName>
    <SecureQuestion>Where were you born?</SecureQuestion>
    <SecureAnswer>Krypton</SecureAnswer>
    <Addresses>
      <Address type="home">
        <Street>638 Shelter Drive</Street>
        <AptNo></AptNo>
        <PostalCode>92478</PostalCode>
        <City>Smallville</City>
        <Region>CA</Region>
        <Country>USA</Country>
      </Address>
    </Addresses>
    <Phones>
      <Phone type="home">609 CRY HELP</Phone>
    </Phones>
    <Emails>
      <Email type="home">superman@heroes.com</Email>
    </Emails>
    <Preferences>
      <Preference type="accommodation">
        <Description>Hotel preferences</Description>
        <PriceRange currency="Dollar">
```

```
      <Low>0</Low>
      <High>50</High>
    </PriceRange>
    <Types>
      <Type prefLevel="5">Hotel</Type>
      <Type prefLevel="4">Motel</Type>
      <Type prefLevel="3">Apartment</Type>
      <Type prefLevel="1">Hostel</Type>
    </Types>
    <Features>
    <Feature prefLevel="1">HBO</Feature>
      <Feature prefLevel="1">Breakfast</Feature>
    </Features>
  </Preference>
  <Preference type="entertainnment">
    <Description>Park preferences</Description>
    <PriceRange currency="Dollar">
      <Low>0</Low>
      <High>75</High>
    </PriceRange>
    <Types>
      <Type prefLevel="4">Waterpark</Type>
      <Type prefLevel="1">Familypark</Type>
      <Type prefLevel="5">Rollercoaster</Type>
      <Type prefLevel="2">Zoo</Type>
      <Type prefLevel="3">Museum</Type>
      <Type prefLevel="4">Live Show</Type>
    </Types>
    <Features>
      <Feature prefLevel="1">Outdoor</Feature>
      <Feature prefLevel="5">Indoor</Feature>
    </Features>
  </Preference>
  <Preference type="restaurant">
    <Description>Restaurant preferences</Description>
    <PriceRange currency="Dollar">
      <Low>0</Low>
      <High>50</High>
    </PriceRange>
    <Types>
      <Type prefLevel="2">African</Type>
      <Type prefLevel="3">American</Type>
      <Type prefLevel="3">Chinese</Type>
      <Type prefLevel="4">Italian</Type>
```

```
          <!-- MANY MORE TYPES --!>
        </Types>
        <Features>
        </Features>
      </Preference>
    </Preferences>
    <Thresholds>
      <PreferenceLevelCutOff>1</PreferenceLevelCutOff>
    </Thresholds>
  </Static>
  <Dynamic>
    <LocationHistory>
      <Location>
        <Timestamp>2003-01-12 18:19:23 GMT-06:00</Timestamp>
        <Latitude unit="dec-deg">42.076111</Latitude>
        <Longitude unit="dec-deg">-87.74888</Longitude>
      </Location>
      <Location>
        <Timestamp>2003-01-12 18:20:23 GMT-06:00</Timestamp>
        <Latitude unit="dec-deg">42.094722</Latitude>
        <Longitude unit="dec-deg">-87.74888</Longitude>
      </Location>
    </LocationHistory>
    <Speed unit="mph">1.42</Speed>
    <Direction unit="dec-deg">63.25</Direction>
    <LocalTime>2003-02-12 18:24:34 GMT-06:00</LocalTime>
    <ActualState>pedestrian</ActualState>
    <UserDevice>pda</UserDevice>
  </Dynamic>
</Profile>
```

Listing 11.10 Context and Profile Manager XML user profile

The user profile contains static preferences within the *<Static>* XML element. These preferences include user identification information such as name, password, addresses, phones, and emails. The user can specify separate preferences for accommodations, entertainment, and restaurants. For accommodations, the user can specify the type of hotel, price range, and any particular features that he or she is looking for, such as included breakfast. For entertainment, the user can rank order the type of parks and attractions that he or she would want to visit. Finally, for restaurants, the user can specify their favourite cuisines, and order these in separate preference levels. The *<Thresholds>* element specifies the preference level below which items, such as restaurants, will not be displayed.

Dynamic elements of context are contained within the *<Dynamic>* XML element. The dynamic elements include a record of the last locations where the user traveled. The user's mobile terminal sends to the CPM location updates. These are sent on a periodic basis and take the form of GPS coordinates, with latitude and longitude degrees. To each location update, the CPM adds a timestamp of when the update was received. These timestamps enable the computation of an average speed of travel. From the location history, the CPM is also able to compute an average direction of travel in degrees, where North is at 0° and degrees are measured counterclockwise to $+180^\circ$ and clockwise to -180°.

Other dynamic elements include the local time at the user's locale, the user's transportation state, that is, whether a pedestrian, a driver, or a passenger, and finally, the type of mobile terminal used. Local time is sent by the user's terminal, and is used to filter Web content, for example, to filter out those restaurants that are presently closed. Transportation state is manually indicated by the user, and the type of mobile terminal is derived from the browser type specified in the HTTP request header.

11.4.3 Context and profile manager filtering

The CPM requests from the restaurant Web service all the restaurants that are included in the user's relevant geographic service area. After retrieving this list, the CPM filters out those restaurants that don't match the user's specified preferences as stated in his or her profile. In addition, the CPM augments the restaurant information with dynamic context information. This context includes the user's distance to the restaurant, the direction from the user to the restaurant, and the wait time for a table (when available). Listing 11.11 shows the CPM's restaurant filtering steps.

```
while( iter.hasNext() )
{
  Restaurant rest = (Restaurant)iter.next();
  //include only restaurants that are currently open
  if(!checkForOpenCloseHours(rest, time) )
  {
    continue;
  }

  //calculate the distance and angle between the restaurant
  //and the user and add this info to the restaurant
  addDistanceAndDirection(rest, latitude, longitude,
    prof.getDirection() );

  //include only restaurants that are reachable within 1/2 hour
  speed = prof.getSpeed();
  if(rest.getDistance() > speed/2 )
```

```
  {
    continue;
  }

  //check if restaurant's price range
  //falls within the user's preferred price range
  if(restaurantPref.getPriceRange().getLow() > rest.getPriceHigh() ||
    restaurantPref.getPriceRange().getHigh() < rest.getPriceLow() )
  {
    continue;
  }
  //add the cuisine's preference level to the restaurant
  //if the cuisine's preference level is above the cutoff
  boolean skip = false;
  boolean foundCuisine = false;
  ContentType[] cuisinesPref = restaurantPref.getContentTypes();
  for(int k=0; k < cuisinesPref.length; k++ )
  {
  if( cuisinesPref[k].getTypeDescription().equals(rest.getCuisine() )
  {
    if( cuisinesPref[k].getPrefLevel() <=
      prof.getThresholds().getPrefLevelCutOff())
    {
      skip = true;
    }else
    {
      rest.setPreferenceLevel( cuisinesPref[k].getPrefLevel() );
      foundCuisine = true;
    }
      break;
    }
  }
  if(skip || !foundCuisine )
  {
    continue;
  }
  //add restaurant to list of returned restaurants
}
```

Listing 11.11 Context and Profile Manager information filtering

The CPM loops through the restaurants and performs the following checks:

- The first *if* statement makes sure that a restaurant is not included if it isn't open yet or is already closed. *time* is the local user's time obtained from the user's profile.

checkForOpenCloseHours obtains the open and close hours from *rest*, the current restaurant object, and compares them to *time*. It returns *true* if *time* is within the open and close hours.

- The call to *addDistanceAndAngleTags* calculates the distance and the direction from the user's location to the restaurant and adds this information to the restaurant object. *rest* is the restaurant object, *latitude* and *longitude* are the GPS coordinates of the user's last location obtained from the user's profile, and *prof.getDirection()* retrieves the user's current direction of travel from the profile.
- The next *if* statement makes sure that only restaurants that are reachable within the next half hour, at the current speed, are included.
- The next *if* statement checks the restaurant's price range. If the restaurant's price range does not have an overlap with the user's specified price range, it is not included.
- The *for* loop goes through all the user's preferred cuisines and compares them with the restaurant's cuisine. If a match is found, the user's cuisine preference level is set in the restaurant object. However, it is only set if the level is above a certain threshold.

Listing 11.12 shows an example XML element of a restaurant that the CPM selected for display to the mobile user. This is the example from Listing 11.9 augmented with four XML elements of user preferences and dynamic context data: <CuisinePreferenceLevel>, <Distance>, <Direction>, and <WaitTime>.

```
<Restaurant>
  <Name>Tanellis Chicago Style Pizza</Name>
  <Street>5807 W Diversey Ave</Street>
  <City>Chicago</City>
  <State>IL</State>
  <Zip>60639</Zip>
  <Phone>(773) 622-4244</Phone>
  <Cuisine>Pizza</Cuisine>
  <Latitude>41.931063</Latitude>
  <Longitude>-87.771360</Longitude>
  <YahooMap>http://imgi.maps.yahoo.com/mapimage?MAPData=ccjS
    Uvhyzy286JaOg23zy23YX1j.Mou6fry7xGSzvdH4a5X3yemK5XBI0j2oJ
    q37aqcygoVqwG_oxBs6dSdG51NsqQOa_Klbjeu21a4z7lwIXRGEufOFu.
    jZTJclveox</YahooMap>
  <Rating>
    <Overall>-</Overall>
    <Food>-</Food>
    <Service>-</Service>
    <Ambience>-</Ambience>
  </Rating>
  <PriceLow>$10</PriceLow>
  <PriceHigh>$18</PriceHigh>
  <Open>9:00</Open>
```

```
    <Close>22:30</Close>
    <Valet>0</Valet>
    <CuisinePreferenceLevel>3</CuisinePreferenceLevel>
    <Distance>1.33</Distance>
    <Direction>76.25</Direction>
    <WaitTime>20</WaitTime>
</Restaurant>
```

Listing 11.12 Restaurant data with augmented dynamic context

11.4.4 Application server transformation

The list of restaurants returned by the CPM to the Application server is in the form of an XML file. This file needs to be transformed into markup appropriate for display on the user's mobile terminal. The transformation is performed by an XSLT processor; the inputs to the processor consist of the XML document and a corresponding XSLT style sheet that is appropriate for the user's mobile terminal. Listing 11.13 shows an XSLT style sheet that converts the list of restaurants into HTML to be displayed on a PDA-type terminal.

```
<xsl:stylesheet xmlns:xsl="http://www.w3.org/1999/XSL/Transform"
 version="1.0">
  <xsl:output method="xml" omit-xml-declaration="yes"/>
  <xsl:template match="/">

  <xsl:variable name="listInfo" select="'LIST_INFO_VARIABLE'"/>

  <TABLE cellspacing="3" cellpadding="8">
    <TR bgcolor="#1f70b0">
      <TD colspan="2">
        <font color="white">
          <B>Restaurant Information</B>
        </font>
      </TD>
    </TR>
    <TR>
      <TD align="left" colspan="2">
        <A HREF="LINK_TO_PREVIOUS">Previous</A> |
        <A HREF="restaurant.jsp?Index=1">Show all restaurants</A> |
        <A HREF="LINK_TO_NEXT">Next</A>
      </TD>
    </TR>
    <TR>
      <TD colspan="2" align="center">
        <b><xsl:value-of select="$listInfo" /></b>
```

```
    </TD>
  </TR>
</TABLE>

<TABLE cellSpacing="2" cellPadding="1" align="center" border="0">
  <xsl:for-each select="Restaurants/Restaurant">
  <xsl:sort select="CuisinePreferenceLevel" order="descending"
    data-type="number"/>
  <xsl:sort select="Distance"order="ascending"data-type="number"/>
  <xsl:sort select="WaitTime"order="ascending"data-type="number"/>
  <xsl:if test="((position() &gt; START_INDEX) or (position() =
   START_INDEX)) and position() &lt; STOP_INDEX">
  <TR>
    <TD><B>Name:</B></TD>
    <TD>
      <xsl:variable name="restName" select="Name"/>
      <A HREF="restaurant.jsp?Detail=1&amp;Name={$restName}&amp;
       Index=1">
        <xsl:value-of select="Name"/>
      </A>
    </TD>
  </TR>
  <TR>
    <TD><B>Cuisine:</B></TD>
    <TD>
      <xsl:variable name="cuisineName" select="Cuisine"/>
      <A HREF="restaurant.jsp?Cuisine={$cuisineName}&amp;Index=1">
        <xsl:value-of select="Cuisine"/>
      </A>
    </TD>
  <TR>
  <TR>
    <TD><B>Distance:</B></TD>
    <TD><xsl:value-of select="Distance" /> miles</TD>
  </TR>
  <TR>
    <TD><B>Direction:</B></TD>
    <TD>
      <xsl:if test="number(Direction) &lt; 0">
        <xsl:choose>
          <xsl:when test="number(Direction) &gt;
            -22.5">N</xsl:when>
          <xsl:when test="number(Direction) &lt;= -22.5 and
           number(Direction) &gt; -67.5">NE</xsl:when>
```

```
            <xsl:when test="number(Direction) &lt;= -67.5 and
             number(Direction) &gt; -112.5">E</xsl:when>
            <xsl:when test="number(Direction) &lt;= -112.5 and
             number(Direction) &gt;-157.5">SE</xsl:when>
            <xsl:when test="number(Direction) &lt;=
              -157.5">S</xsl:when>
          </xsl:choose>
        </xsl:if>
        <xsl:if test="number(Direction) &gt;= 0">
          <xsl:choose>
            <xsl:when test="number(Direction) &lt;
             22.5">N</xsl:when>
            <xsl:when test="number(Direction) &gt;= 22.5 and
             number(Direction) &lt; 67.5">NW</xsl:when>
            <xsl:when test="number(Direction) &gt;= 67.5 and
             number(Direction) &lt; 112.5">W</xsl:when>
            <xsl:when test="number(Direction) &gt;= 112.5 and
             number(Direction) &lt; 157.5">SW</xsl:when>
            <xsl:when test="number(Direction) &gt;=
              157.5">S</xsl:when>
          </xsl:choose>
        </xsl:if>
      </TD>
    </TR>
    <TR><TD colspan="2"><HR /></TD></TR>
    </xsl:if>
    </xsl:for-each>

    <TR><TD colspan="2"><BR /></TD></TR>
    <TR><TD colspan="2"><BR /></TD></TR>
    <TR>
      <TD align="left" colspan="2">
        <A HREF="LINK_TO_PREVIOUS">Previous</A> |
        <A HREF="restaurant.jsp?Index=1">Show all restaurants</A> |
        <A HREF="LINK_TO_NEXT">Next</A>
      </TD>
    </TR>
  </TABLE>
  <HR />

  </xsl:template>
</xsl:stylesheet>
```

Listing 11.13 XSLT style sheet for restaurant list in HTML

The *<sort>* elements order the list of displayed restaurants first by the user's pre-specified cuisine preference level, then by increasing distance from the user, and finally by wait time for an available table. The restaurant's name is displayed as a hypertext link to allow the user to request more detailed information about the restaurant. The *<choose>* element provides multiple choices for selecting the acronym ("N", "NE", etc) of the direction to the restaurant based on the direction's angle as specified in the restaurant's *<Direction>* element.

An example of a list of restaurants that is displayed on a PDA-type terminal as a result of the XSLT transformation is shown in Figure 11.1. After the user selects one of the restaurants in the list, the Application server will send the detailed restaurant information to the user's terminal. This information is originally in XML form and needs to be transformed into the appropriate markup. The XSLT style sheet in Listing 11.14 generates HTML markup for detailed restaurant information on a PDA-type terminal.

```
<xsl:stylesheet xmlns:xsl="http://www.w3.org/1999/XSL/Transform"
 version="1.0">
  <xsl:output method="xml" omit-xml-declaration="yes"/>
  <xsl:template match="/">

  <TABLE cellspacing="3" cellpadding="8">
  <TR bgcolor="#1f70b0">
    <TD colspan="2">
      <font color="white">
        <B>Restaurant Information</B>
      </font>
    </TD>
  </TR>
  <TR>
    <TD align="left" colspan="2">
      <A HREF="restaurant.jsp?Index=1">Show all restaurants</A> |
    </TD>
  </TR>
  </TABLE>

  <TABLE cellSpacing="2" cellPadding="1" align="center" border="0">
  <xsl:for-each select="Restaurants/Restaurant [starts-with(Name,
    'SPECIFIC_NAME')]">
  <TR>
    <TD><B>Name:</B></TD>
    <TD>
      <xsl:value-of select="Name"/>
    </TD>
```

```
</TR>
<TR>
  <TD><B>Address:</B></TD>
  <TD>
    <xsl:value-of select="Street" /><BR/>
    <xsl:value-of select="City" />,
    <xsl:value-of select="concat(State, '', Zip)" />
  </TD>
</TR>
<TR>
  <TD><B>Phone:</B></TD>
  <TD><xsl:value-of select="Phone" /></TD>
</TR>
<TR>
  <TD><B>Distance:</B></TD>
  <TD><xsl:value-of select="Distance" /> miles</TD>
</TR>
<TR>
  <TD><B>Direction:</B></TD>
  <TD>
    <xsl:if test="number(Direction) &lt; 0">
      <xsl:choose>
        <xsl:when test="number(Direction) &gt;-22.5">N</xsl:when>
        <xsl:when test="number(Direction) &lt;= -22.5 and
          number(Direction) &gt;-67.5">NE</xsl:when>
        <xsl:when test="number(Direction) &lt;= -67.5 and
          number(Direction) &gt;-112.5">E</xsl:when>
        <xsl:when test="number(Direction) &lt;= -112.5 and
          number(Direction) &gt;-157.5">SE</xsl:when>
        <xsl:when test="number(Direction)&lt;=-157.5">S</xsl:when>
        </xsl:choose>
        </xsl:if>
        <xsl:if test="number(Direction) &gt;= 0">
        <xsl:choose>
        <xsl:whentest="number(Direction)&lt;22.5">N</xsl:when>
        <xsl:when test="number(Direction) &gt;= 22.5 and
          number(Direction) &lt; 67.5">NW</xsl:when>
        <xsl:when test="number(Direction) &gt;= 67.5 and
          number(Direction) &lt; 112.5">W</xsl:when>
        <xsl:when test="number(Direction) &gt;= 112.5 and
          number(Direction) &lt;157.5">SW</xsl:when>
        <xsl:when test="number(Direction) &gt;=157.5">S</xsl:when>
      </xsl:choose>
```

```
    </xsl:if>
  </TD>
</TR>
<TR>
  <TD><B>Cuisine:</B></TD>
  <TD>
    <xsl:variable name="cuisineName" select="Cuisine"/>
    <A HREF="restaurant.jsp?Cuisine={$cuisineName}&amp;Index=1">
      <xsl:value-of select="Cuisine"/>
    </A>
  </TD>
</TR>
<TR>
  <TD><B>Open:</B></TD>
  <TD>
    <xsl:value-of select="Open"/> -
    <xsl:value-of select="Close"/>
  </TD>
</TR>
<TR>
  <TD><B>Price:</B></TD>
  <TD>
    <xsl:value-of select="PriceLow"/> -
    <xsl:value-of select="PriceHigh"/>
  </TD>
</TR>
<TR>
  <TD><B>Waittime:</B></TD>
  <TD><xsl:value-of select="Waittime" /></TD>
</TR>
<TR>
  <TD><B>Valet:</B></TD>
  <TD>
    <xsl:if test="contains(Valet,'0')">No</xsl:if>
    <xsl:if test="contains(Valet,'1')">Yes</xsl:if>
  </TD>
</TR>
<TR>
  <TD><B>Map:</B></TD>
  <TD><A HREF="{YahooMap}" target="_blank">Location map</A></TD>
</TR>
<TR><TD colspan="2"><HR /></TD></TR>
</xsl:for-each>
```

```
  <TR><TD colspan="2"><BR /></TD></TR>
  <TR><TD colspan="2"><BR /></TD></TR>
  <TR>
    <TD align="left" colspan="2">
      <A HREF="restaurant.jsp?Index=1">Show all restaurants</A> |
    </TD>
  </TR>
  </TABLE>
  <HR />

  </xsl:template>
</xsl:stylesheet>
```

Listing 11.14 XSLT style sheet for restaurant detail in HTML

REFERENCES AND FURTHER READING

[1] R. Blättler, Context-aware information system for mobile users. M.S. Project, Northwestern University (Jun. 2002).

[2] A. Heusser, Context-aware information system for mobile users, part II. M.S. Project, Northwestern University (Jun. 2003).

[3] A. Pashtan *et al.*, CATIS: A Context-Aware Tourist Information System. *Proc. of the 4th International Workshop of Mobile Computing*, Rostock, June 2003.

[4] Apache Xindice database, http://xml.apache.org/xindice/.

[5] K. Staken, *An Introduction to the XML:DB API*, http://www.xml.com/pub/a/2002/01/09/xmldb_api.html (Jan. 2002).

Glossary

2.5G	Second and a Half Generation
3G	Third Generation
3GPP	3rd Generation Partnership Project
3GPP2	Third Generation Partnership Project 2
AAA	authentication, authorization and accounting
A-GPS	Assisted GPS
ANSI	American National Standards Institute
ARPU	average revenue per user
ASP	Application Service Provider
CATIS	Context-Aware Tourist Information System
CC/PP	Composite Capabilities/Preference Profiles
CC/PPEX	CC/PP Exchange Protocol
CCQ	Client Capabilities Query
CDATA	XML type character data
CDMA	Code Division Multiple Access
CGI	Common Gateway Interface
cHTML	Compact Hypertext Markup Language
CPM	Context and Profile Manager
CSS	Cascading Style Sheet
DAML	DARPA Agent Markup Language
DARPA	Defense Advanced Research Projects Agency
DBMS	Database Management System
DES	Data Encryption Standard
D-GPS	Differential GPS
DHCP	Dynamic Host Configuration Protocol
DNS	domain name server
DOM	Document Object Model
DSSSL	Document Style and Semantics Specification Language
DTD	Document Type Definition
EDGE	Enhanced Data rates for GSM Evolution

EDI	Electronic Data Interchange
EJB	Enterprise JavaBean
E-OTD	Enhanced Observed Time Difference
FCC	US Federal Communications Commission
FDD	Frequency Division Duplex
FIPA	Foundation for Intelligent Physical Agents
GGSN	gateway GPRS support node
GIF	Graphics Interchange Format
GIS	geographic information system
GMLC	Gateway Mobile Location Center
GPRS	General Packet Radio Service
GPS	Global Positioning System
GSM	Global System for Mobile communication
GSMA	Global System for Mobile Communications Association
HDML	Handheld Device Markup Language
HLR	Home Location Register
HTML	Hypertext Markup Language
HTTP	Hypertext Transfer Protocol
HTTPS	Secure HTTP
IETF	Internet Engineering Task Force
IMSI	International Mobile Subscriber Identity
IMT	International Mobile Telecommunications
IP	Internet Protocol
IR	infrared
ISO	International Organization for Standardization
ISP	Internet Service Provider
ITU	International Telecommunication Union
ITU-T	ITU Telecommunication Standardization Sector
J2EE	Java 2 Enterprise Edition (TM?)
J2ME	Java 2 Micro-Edition
JAXM	Java API for XML Messaging
JAX-RPC	Java API for XML RPC
JDBC	Java Database Connectivity
JSP	Java Server Pages (TM?)
kbps	kilobits per second
LDAP	lightweight directory access protocol
LIF	Location Interoperability Forum
LMU	location measurement unit
LSC	location service center
MBR	Minimum Bounding Rectangle

MIME	Multipurpose Internet Mail Extensions
MLP	Mobile Location Protocol
MOD	Moving objects database
MP	Mobile Profile
NAI	network access identifier
NAT	network address translator
NIST	National Institute of Standards and Technology
NMTOKEN	XML type name token
OASIS	Organization for the Advancement of Structured Information Standards
OIL	Ontology Inference Layer
OMA	Open Mobile Alliance
OS	Operating system
OSA	Open Service Access
OTDOA	Observed Time Difference of Arrival
OWL DL	OWL Description Logics
OWL	Web Ontology Language
OWL-S	OWL Service Ontology
P3P	Platform for Privacy Preferences
PAP	Push Access Protocol
PCDATA	XML type parsed character data
PDA	Personal Digital Appliance
PDC-P	Personal Digital Cellular – Packet
PI	Push Initiator
PIN	Personal Identification Number
PKCS	public-key cryptography standards
PKI	public key infrastructure
POI	Point of interest
PPG	Push Proxy Gateway
PSAP	Public Safety Answering Point
PTA	Personal Travel Assistance
RDF	Resource Description Framework
RDFS	Resource Description Framework Schema
RFC	Request for comment
RPC	Remote Procedure Call
RTT	Radio Transmission Technology
SA	Selective Ability
SAML	Security Assertion Markup Language
SDK	Software Development Kit
SGML	Standard Generalized Markup Language
SGSN	serving GPRS support node

SI	service indication
SIR	Session Initiation Request
SL	service loading
SMIL	Synchronized Multimedia Integration Language
SMLC	Serving Mobile Location Center
SMS	Short Message Service
SNOMED	Systemized Nomenclature of Medicine
SOAP	Simple Object Access Protocol
SQL	Structured Query Language
SSL	Secure Socket Layer
SSO	single sign-on
TA	Timing Advance
TCP	Transmission Control Protocol
TCP/IP	Transmission Control Protocol/Internet Protocol
TDD	Time Division Duplex
TDMA	Time Division Multiple Access
TL	Transport layer
TLS	Transport Layer Security
UDDI	Universal Description, Discovery and Integration
URI	Universal Resource Indicator
URL	Universal Resource Locator
USB	Universal Serial Bus
U-TDOA	Uplink Time Difference of Arrival
UTRA	Universal Terrestrial Radio Access
VLR	Visitor Location Register
W3C	World Wide Web Consortium
WAE	WAP Application Environment
WAP	Wireless Application Protocol
WBMP	Wireless Bit Map
WCSS	wireless CSS or WAP CSS
WDP	Wireless Datagram Protocol
Web	World Wide Web
WG	W3C working group
WLAN	wireless local area network
WML	wireless markup language
WP	wireless profiled
WP-HTTP	wireless-profiled HTTP
WP-TCP	wireless-profiled TCP
WS	Web Service
WSDL	Web Services Description Language

WSP	Wireless Session Protocol
WTA	Wireless Telephony API
WTLS	Wireless Transport Layer Security
WTP	Wireless Transaction Protocol
WWAN	wireless wide area network
XHTML	Extensible Hypertext Markup Language
X-KISS	XML Key Information Service Specification
XKMS	XML Key Management Specification
X-KRSS	XML Key Registration Service Specification
XML	Extensible Markup Language
XML-DB	XML Database
XSL	Extensible Style Sheet Language
XSLT	XSL Transformation

Index